Mathematical Wizardry
for a Gardner

Mathematical Wizardry for a Gardner

Edited by
Ed Pegg Jr.
Alan H. Schoen
Tom Rodgers

CRC Press
Taylor & Francis Group
Boca Raton London New York

CRC Press is an imprint of the
Taylor & Francis Group, an **informa** business

AN A K PETERS BOOK

CRC Press
Taylor & Francis Group
6000 Broken Sound Parkway NW, Suite 300
Boca Raton, FL 33487-2742

First issued in paperback 2017

ISBN 13: 978-1-138-11671-9 (pbk)
ISBN 13: 978-1-56881-447-6 (hbk)

Visit the Taylor & Francis Web site at
http://www.taylorandfrancis.com

and the CRC Press Web site at
http://www.crcpress.com

"Martin Gardner and Paperfolding" copyright © David Lister.

Cover illustrations and design by Victoria E. Kichuk.

Library of Congress Cataloging-in-Publication Data

Mathematical wizardry for a Gardner / edited by Ed Pegg Jr., Alan H. Schoen, [and] Tom Rodgers.
    p. cm.
  Includes bibliographical references.
  ISBN 978-1-56881-447-6 (alk. paper)
  1. Mathematical recreations. 2. Gardner, Martin, 1914- I. Gardner, Martin, 1914– II. Pegg , Ed, 1963– III. Schoen, Alan H. (Alan Hugh), 1924– IV. Rodgers, Tom, 1943–
    QA95.M3686 2008
    793.74–dc22

                                                        2008032834

Contents

Contents

Contents

Preface

Mathematicians, puzzlers, and magicians assemble in Atlanta every two years or so to pay tribute to Martin Gardner, whose lifelong devotion to recreational mathematics and to magic is honored throughout the world. The seventh of these "Gatherings for Gardner," G4G7, was held March 16–19, 2006. Most of the articles in this volume, which is a companion to the recent *Homage to a Pied Puzzler*, are drawn from oral presentations delivered at G4G7.

Mathematicians everywhere recognize that even though Martin Gardner is not himself a professional mathematician, he has been astoundingly effective in popularizing not just recreational mathematics, magic, and puzzles, but also so-called serious mathematics. A number of professional mathematicians have acknowledged that their passion for their subject was first kindled by one of Martin's columns or essays concerning either a hoary old chestnut or a brand-new problem. His columns in *Scientific American*, which first appeared in 1956, earned him readers worldwide, and he continues writing to this day. His published writings are available in both book form and on CD.

Martin's unique influence has been the result of a refined taste in subject matter combined with a famously clear and witty writing style. His well-known addictions to magic and to word puzzles have inspired countless others to attempt to emulate him in these arts, but he is still the undisputed master. Perhaps his most brilliant invention is Dr. Irving Joshua Matrix, a proxy he used to spoof virtually every known variety of charlatan, including—but not limited to—Biblical cryptographers and numerologists.

The first two articles in this book are dedicated to the memory of Frank Harary, mathematician *extraordinaire*, who died early in 2005. Frank was a welcome presence at G4G6, where—in spite of his somewhat frail health—he succeeded in brightening the days of the gatherers, especially with his affectionately warm wit. Those of us who were lucky enough to count Frank as a friend sorely miss him.

We also mourn the passing in 2008 of the breathtakingly encyclopedic polymath Steve Sigur, whose contributions to geometry are legendary. Steve, who participated in G4G8 just three months before his death, was widely celebrated not only for the range and depth of his knowledge, but also for his unusual generosity to his students.

It was Tom Rodgers who in 1993 first thought of assembling admirers of Martin Gardner for a few days of celebration. He has hosted all of these gatherings since then, with the generous help of Elwyn Berlekamp, Karen Farrell, Jeremiah Farrell, Scott Hudson, Emily DeWitt Rodgers, Mark Setteducati, Stephen Turner, Thane Plambeck, and the many G4G participants who volunteered at moments of need.

The editors are especially indebted to Charlotte Henderson, associate editor for our publisher, A K Peters, Ltd., for her patient ministrations in the task of assembling and editing authors' contributions.

We hope that readers derive as much pleasure from this volume and its companion volume as we have had in assembling them.

Ed Pegg Jr.
Champaign-Urbana, Illinois

Alan H. Schoen
Carbondale, Illinois

Tom Rodgers
Atlanta, Georgia

In Memoriam

Frank Harary

Gary Chartrand

I met Frank Harary at the University of Michigan in October 1963 while I was a graduate student at Michigan State University. This visit led to my being invited to spend the academic year 1965–1966 to assist him while he was working on his classic textbook, *Graph Theory*. I always felt, however, that he assisted me much more than any help I may have given him. I saw how he did research and I paid careful attention to the way he thought about and analyzed mathematical problems. We became and remained friends until the day he died. Although he ordinarily closed his emails with "Best, Frank," he would often end his emails to me with "Your friend, Frank." He was not only a friend to me; he became my mentor. I sought his advice on many topics and I valued his opinions. It seemed like whatever topic I discussed with him, he had previously considered it and developed an opinion on it.

Frank had so many interests. He was a connoisseur of "good" wine, he loved writing poetry, and he was an avid investor in the stock market. I once asked him if he really understood what he was doing when he invested in stocks, to which he answered "Not at all but I enjoy it." He said that he had become a millionaire on three occasions but each time lost much of the money. He said that if being wealthy was important to him, he would marry one of his ex-wives.

It was impossible to turn down anything he would ask me to do. Such a request would often start with "Could you do me a favor?" In the fall of 1974, I received a phone call at home one evening from Frank, who told me that he had been in discussions with Wiley about developing a new journal that he was going to call the *Journal of Graph Theory.* He said, however, that he wouldn't go ahead with his plans unless I agreed to be the managing editor. Not that I really believed that but I remember telling him something like: Frank, I'm teaching two classes, doing research, writing a book, and directing doctoral dissertations. I don't have the time but, definitely, the answer is yes.

When I was the managing editor of *JGT,* I would often drive to Ann Arbor to discuss the journal with him. At times, he would meet me midway at a restaurant in Marshall, Michigan. It was during these meetings that I really began to see and understand his philosophy about research. He firmly believed that the journal should contain and emphasize articles that dealt with BGT and not just GT. He would write and say "BGT" for "Beautiful Graph Theory." He would often say, "Just because it's true doesn't mean you have to publish it."

He felt that research should concentrate on pretty concepts and results, topics that would be fun and interesting to lecture on and to listen to. Frank had such a fertile mind. I recall attending a graph theory conference one Saturday at the University of Michigan where after each talk, Frank would ask probing questions and suggest new problems for the speaker to work on.

I remember saying to Frank that what constitutes interesting mathematics is so subjective. His response to this was, "That's why we need to have good taste." He said that doing good mathematics relies on being able to ask interesting questions and to ask the right questions. You only have so much time in your life and you have to make decisions on how you intend to spend that time. Even if you're trying to prove a difficult result and you're successful but the result is boring, then what have you accomplished? So if you're faced with a result that you're trying to prove and it appears that it will be a time-consuming activity to prove it, then you have to compare the time you would spend trying to prove the result against the value of knowing that the result is true to see if this would be a wise use of your time. Frank would say that such decisions, although on a very small scale, are not unlike decisions that scientists and other scholars must make on a regular basis.

Even nonscholars are faced with making these kinds of decisions, Frank would also say. A world leader has a major decision to make, such as should the country go to war. What are the ramifications of going to war or not going to war? And if the country does go to war, then what are the consequences—politically, in terms of human life, financially, the country's reputation in the world? What are the things that can (and probably will) go wrong and how will they be dealt with?

Frank and I shared a love of music. Although his tastes were more on the classical side and mine were for musical theater, he enjoyed musical theater as well and would often send me programs of shows and concerts he attended. We discussed the fact that so many mathematicians enjoy music. There seemed to be some similarity between creating mathematics and creating (composing) music. Music has the edge, he would say, because it affects the emotions, while mathematics usually does not. Also, good music is appreciated by large numbers of people, while even good mathematics is appreciated by relatively few. When a mathematical topic is being investigated and theorems are being discovered and proved, in effect a mathematical story is being told. In musical theater, there is also a story but songs are added. These songs provide an emotional or comical way to tell the audience either what someone is thinking or else to describe the personality of the character. The closest thing to this in mathematics is when one is talking about what has been discovered. Then humor and emotion can (and probably should) play a role.

For Frank, it was not only important that the mathematics be interesting, it was essential that it be presented (orally or in writing) in a clear and interesting manner. For example, although there are grammatical rules for where commas and quotation marks should be placed (should quotation marks follow or precede a period?), he felt such decisions should be based on common sense. He appreciated the care that Paul Halmos, Donald Knuth, and other scholars took with their writing. Frank was annoyed when he didn't understand the mathematics he was reading. When someone then explained it to him, he would say, "So that's what he meant. Then why didn't he say so?"

He often mentioned that garbled writing was commonly the result of people trying to say too much. He felt that it was always better to keep things short and simple. His poetic reaction to this was:

If confusion runs rampant in the passage just read
It may very well be that too much has been said.

No matter how clearly I felt that I had written mathematics, Frank always found a way to make it even clearer. When I apologized to him for not doing a better job, he came back with one of his most commonly used phrases: "Not to worry."

When things were not going well for me, I would invariably contact Frank, as I knew he would be sympathetic and would offer good advice. He would always respond by saying that things had gone wrong often in his life. What he always did was turn to graph theory because he knew that he could count on it. Frank had a great love for graph theory. He thought it was important for others to know how it developed and to be familiar with the people responsible for this. He felt that the *Journal of Graph Theory* could be used to disseminate this information. In his lectures, Frank would often tell stories about mathematicians. In a lecture where I was present, he once referred to the "top ten graph theorists." At the end of his talk, a member of the audience asked, "Who decides who the top ten graph theorists are anyway?" Without a moment of hesitation, Frank responded, "The ten of us!" In summary then, wherever Frank Harary may fit in among the top graph theorists, one thing remains indisputable: There was and will only be one graph theorist like Frank Harary.

Harary

Jeremiah Farrell

The search for truth is the noblest occupation of man,
its publication a duty.

—Madame de Staël

"Here is another old poser," he offered with a twinkle in his eye,
"Arrange the ten digits across two rows so that the sum of all of
the digits in each row has the same total."

We were at the sixth Gathering for Gardner and, as was our
habit, we were having a private luncheon so that Frank Harary
could bring me up to date on his latest endeavors. I was always
a very willing listener and anxiously tried to solve his "old poser"
that was nevertheless new to me. I will give the reader my solution
(arrived at some time later) at the end of this tribute.

Frank Harary, died at 83 in Las Cruces, NM, on January 4, 2005 after a brief illness. Dr. Harary was widely
recognized as the "father" of modern graph theory, a dis-
cipline of mathematics he helped found, popularize, and
revitalize. [2]

I first met Frank Harary some fifteen years ago when ten of my
students, on a very cold, dreary fall day, braved my driving them in

a rented van to hear him lecture at Indiana University, Kokomo. We knew all about Harary, having read his 1969 classic book *Graph Theory* [6] and we much anticipated his talk. We were not disappointed. In walked the dapper Harary with his perfect posture and short, deliberate steps, and he regaled us for an hour with a brilliant introduction to Fatty, Skinny, Knobby, etc., his names for the "animals" or polyominoes on which he played avoidance and achievement games. Martin Gardner's fine expository article "Generalized Ticktacktoe" [3] explores the deep recreational aspects of this topic. Harary's lecture was filled with humor. I remember especially his opening self-deprecating joke. "No matter which way I turn, I am always facing the board." We were definitely not bored.

> Frank had a beautiful spirit! He had a great joie de vivre which overflowed naturally and spontaneously. He had a sweet heart and gentle disposition. He was a free spirit, who walked to the beat of a different drummer. He loved the arts and attended theater and concert performances with great appreciation. He was fun loving, had a great sense of humor and was a delight to be around. Dr. Harary loved to travel all over the globe to spread the gospel of graph theory—he delivered over a thousand conferences and invited lectures in more than 87 countries in four different languages. [8]

Harary wrote, with over 250 coauthors, more than 700 articles and books. I was privileged to be a Harary coauthor along with my former student, Putnam Fellow, and Gathering attendee Christopher Mihelich (who also was first-prize winner in the Westinghouse Science Contest). Our article "The Elucidators" [1] was "Dedicated to that great elucidator Martin Gardner—nulli secundus" and described some new ways of looking at some puzzles common to both Sam Loyd (1841–1911) and Henry Ernest Dudeney (1857–1930). With the publication of this article, Christopher and I inherited an Erdős number of 2 from Harary's Erdős number 1. We were just as pleased to become new holders of "Harary number 1s."

I cannot resist mentioning some of Harary's other, rather outré articles. One he often joked about was "Is the Null Graph a Pointless Concept?" [7] This included, labeled as "Figure 1. The Null Graph," a totally blank diagram! Yet it is not a completely frivolous article but attempts to decide whether the introduction of the null graph results in a simplification of the statement of several theorems. In the end, however, "No conclusion is reached."

Two of his early articles concerned applying directed graphs to fields other than mathematics, something he was always anxious to do. One, "Who Eats Whom?" [4], relates a previous sociological model to a zoological setting. It provides a combinatorial formula for the "status" of the power of each of 15 different animals. The other article, "Cosi Fan Tutte—A Structural Study" [5], uses directed graphs to study the stability of group structures in Mozart's 1790 opera. The peripatetic Harary's affiliation for this article is listed as the Tavistock Institute of Human Relations, London. Three revealing footnotes enhance the article:

1. Performance at Sadler's Wells, London, April 12, 1963.

2. This note was invited by the Editors and was not supported in any way whatsoever.

3. The letters F_1 and F_2 were chosen to indicate females; it is entirely coincidental that F is also the initial letter of "fickle."

Frank Harary was always entertaining and very gracious to everyone. David Dillon, who now is a researcher in computer algorithms, was one of the ten students that accompanied me to Kokomo fifteen years ago and David also attended the sixth Gathering where he was re-introduced to Harary. They started to exchange ideas for joint projects in the months following the Gathering, and when Harary learned that David's three-year-old daughter was confused about his name, he composed for her this charming poem:

For Bella . . .

There goes Harary
In His Ferrari
On a Safari

And signed it "Fondest regards, Frank."

We will all miss Frank and I especially regret that he and I can no longer have our informative luncheon tutorials. Perhaps I have learned enough from him to continue to disseminate some of his ideas by following the dictum of Madame de Staël as he himself always so ably did.

Answer to Poser

My answer to Harary's opening poser is to "cross" the two rows 1,2,5,8,9 and 3,4,5,6,7 at the 5 for a sum of 25 in each row (the 0 can be placed anywhere). There are other solutions. For instance the common sum of 23 may be obtained by crossing at the digit 1, and crossing at 9 can yield the sum 27.

Bibliography

[1] J. Farrell, F. Harary, and C. C. Mihelich. "The Elucidators." *Word Ways: The Journal of Recreational Linguistics* 33:3 (August, 2000), 194–200.

[2] "In Memoriam." *Focus* 25:3 (March, 2005), 21.

[3] Martin Gardner. "Generalized Ticktacktoe." In *Fractal Music, Hypercards and More*, pp. 202–213. New York: W. H. Freeman, 1991.

[4] Frank Harary. "Who Eats Whom?" *General Systems* 6 (1961), 41–44.

[5] Frank Harary. "Cosi Fan Tutte—A Structural Study." *Psychological Reports* 13 (1963), 466.

[6] Frank Harary. *Graph Theory.* Reading, MA: Addison-Wesley, 1969.

[7] Frank Harary and Ronald C. Read. "Is the Null Graph a Pointless Concept?" *Springer Lecture Notes (Math)* 406 (1974), 37–44.

[8] "Obituary." *Las Cruces Sun-Times.* January 7, 2005. (Available at http://www.cs.nmsu.edu/fnh/obit.html.)

Part I

Spin a Tale

The Ig Nobel Prizes

Stanley Eigen

The Ig Nobel Prizes have been given out every year since 1991 by the international science humor magazine *Annals of Improbable Research* or *AIR* as it is more commonly known. The 2005 ceremony was held on October 6th at Harvard University's historic Sanders Theatre. It was attended by a live audience of 1,200 lab coat and sundry attired individuals who continuously flooded the stage with paper airplanes and other ephemera. This was a slight inconvenience as the ceremony's traditional stage sweeper, Harvard physics professor Roy Glauber, was not in attendance. Two days before the Ig Nobel Ceremony, he received a telephone call from Stockholm, informing him that he had been awarded a Nobel Prize in physics. Eight of the ten Ig Nobel Prize winners traveled to Harvard—at their own expense—to accept their prizes, which were handed to them personally by Nobel Laureates Dudley Herschback (Chemistry '86), William Lipscomb (Chemistry '76), Sheldon Glashow (Physics '79), and Robert Wilson (Physics '78). Wilson, by the way, was the prize in the annual Win-a-Date-with-a-

Figure 1. The Ig Nobel prizes: For achievements that first make people *laugh* and then make them *think*. (Courtesy of Marc Abrahams, ® Improbable Research.)

Nobel-Laureate Contest. (A video is available at http://www.improb .com—of the ceremony, not the date.)

A great deal of mathematics and calculations go into winning some of these Gardneresque prizes. Here are a few winners from previous years:

- The Southern Baptist Church of Alabama won for calculating how many Alabama citizens will go to Hell if they don't repent [2]. Answer: 46.1

- Robert Faid of Greenville, South Carolina, won for calculating the exact odds that Mikhail Gorbachev is the Antichrist [4]. Answer: 710,609,175,188,282,000 to 1.

- Arnd Leike of the University of Munich won for demonstrating that the mathematical Law of Exponential Decay applies to beer froth [6].

- Jerald Bain of Mt. Sinai Hospital in Toronto and Kerry Siminoski of the University of Alberta won for their study "The Relationship Among Height, Penile Length, and Foot Size" [1].

- K.P. Sreekumar and the late G. Nirmalan of Kerala Agricultural University, India, won for their "Estimation of the Total Surface Area in Indian Elephants (Elephas maximus indicus)" [12].

- Bernard Vonnegut of the State University of Albany won for "Chicken Plucking as Measure of Tornado Wind Speed" [13]. The Chicken Plucking problem goes back to 1842. The question was, as stated, "to measure the wind speed of tornados". The issue was—at least in part—that the researchers could not get to the scene until after a tornado passed. Fortunately, they were able to identify an analog of sorts—plucked chickens. For comparison purposes, chickens were shot out of a cannon. Professor Vonnegut reviewed the research and concluded it didn't work.

Here are some of the winners from 2005:

- James Watson of Massey University, New Zealand, won the Agricultural History award for his scholarly study, "The Significance of Mr. Richard Buckley's Exploding Trousers" [14].

- John Mainstone and the late Thomas Parnell of the University of Queensland, Australia, won the Physics award for patiently conducting an experiment that began in the year 1927—in which a glob of congealed black tar has been slowly, slowly dripping through a funnel, at a rate of approximately one drop every nine years [3].

- Claire Rind of Newcastle University, in the UK, won the Peace award for electrically monitoring the activity of a brain cell in a locust while that locust was watching selected highlights from the movie *Star Wars* [9].

- Gauri Nanda of the Massachusetts Institute of Technology won the Economics award for inventing an alarm clock that runs away and hides, repeatedly, thus ensuring that people *do* get out of bed, adding many productive hours to the workday. (We saw the clock work.)

- Edward Cussler of the University of Minnesota and Brian Gettelfinger of the University of Minnesota and the University of Wisconsin won the Chemistry award for conducting a careful experiment to settle the longstanding scientific question: can people swim faster in syrup or in water? [5] (If you are like me, you will be disappointed to learn the answer is neither. They go the same speed. By the way, Isaac Newton tried his hand at this problem—Cussler and Gettelfinger showed that Newton got it wrong.)

- Benjamin Smith of the University of Adelaide, Australia, and the University of Toronto, Canada, the Firmenich perfume company, Geneva, Switzerland, and ChemComm Enterprises, Archamps, France, and Craig Williams of James Cook University and the University of South Australia won the Biology award; for painstakingly smelling and cataloging the peculiar odors produced by 131 different species of frogs when the frogs were feeling stressed. The prize was shared with the following who could not attend the ceremony: Michael Tyler of the University of Adelaide, Brian Williams of the University of Adelaide, and Yoji Hayasaka of the Australian Wine Research Institute. [10, 11]

- Dr. Yoshiro Nakamatsu of Tokyo, Japan, won the Nutrition award for photographing and retrospectively analyzing every meal he has consumed during a period of 34 years (and counting). (If you don't know who Dr. Nakamatsu is—well, all I can say is I was delighted to meet him and I recommend you look him up on the web.)

The following winners could not attend but sent in video acceptance speeches:

- Gregg A. Miller of Oak Grove, Missouri, won the Medicine award for inventing Neuticles—artificial replacement testicles for dogs, which are available in three sizes and three degrees of firmness (US Patent #5868140) [8].

- Victor Benno Meyer-Rochow of International University, Bremen, Germany, and the University of Oulu, Finland; and

Jozsef Gal of Lorand Eotvos University, Hungary, won the Fluid Dynamics award for their report "Pressures Produced When Penguins Pooh—Calculations on Avian Defecation" [7].

Bibliography

[1] Jerald Bain and Kerry Siminoski. "The Relationship Among Height, Penile Length, and Foot Size." *Annals of Sex Research* 6:3 (1993), 231–235.

[2] "Baptists Count the Lost 46% of Alabamians Face Damnation, Report Says." *Birmingham News*, September 5, 1993, PAGE.

[3] R. Edgeworth, B. J. Dalton, and T. Parnell. "The Pitch Drop Experiment." *European Journal of Physics* 5:4 (1984), 198–200.

[4] Robert Faid. Gorbachev! Has the Real Antichrist Come? Tulsa: Victory House, 1988.

[5] Brian Gettelfinger and E. L. Cussler. "Will Humans Swim Faster or Slower in Syrup?" *American Institute of Chemical Engineers Journal* 50:11 (2004), 2646–2647.

[6] Arnd Leike. "Demonstration of the Exponential Decay Law Using Beer Froth." *European Journal of Physics* 23:1 (2002), 21–26.

[7] Victor Benno Meyer-Rochow and Jozsef Gal. "Pressures Produced When Penguins Pooh—Calculations on Avian Defecation." *Polar Biology* 27 (2003), 56–58.

[8] Gregg A. Miller. Going ...Going ...NUTS! Frederick, MD: *Publish America*, 2004.

[9] F. C. Rind and P. J. Simmons. "Orthopteran DCMD Neuron: A Reevaluation of Responses to Moving Objects. I. Selective Responses to Approaching Objects." *Journal of Neurophysiology* 68:5 (1992), 1654–1666.

[10] Benjamin P.C. Smith, Michael J. Tyler, Brian D. Williams, and Yoji Hayasaka. "Chemical and Olfactory Characterization of Odorous Compounds and Their Precursors in the Parotoid Gland Secretion of the Green Tree Frog (Litoria caerulea)." *Journal of Chemical Ecology* 29:9 (September 2003), 2085–2100.

[11] Benjamin P.C. Smith, Craig R. Williams, Michael J. Tyler, and Brian D. Williams. "A Survey of Frog Odorous Secretions, Their Possible Functions and Phylogenetic Significance." *Applied Herpetology* 2:1–2 (February 1, 2004), 47–82.

[12] K.P. Sreekumar and G. Nirmalan. "Estimation of the Total Surface Area in Indian Elephants (Elephas maximus indicus)." *Veterinary Research Communications* 14:1 (1990), 5–17.

[13] Bernard Vonnegut. "Chicken Plucking as Measure of Tornado Wind Speed." *Weatherwise* (October 1975), 217.

[14] James Watson. "The Significance of Mr. Richard Buckley's Exploding Trousers." *Agricultural History* 78:3 (Summer 2004), 346–360.

Martin Gardner and Paperfolding

David Lister

Martin Gardner is best known for his long-running monthly column "Mathematical Games" in the venerable science magazine *Scientific American*. As his column developed it embraced many topics beyond the strict interpretation of its title, but even then the column did not by any means exhaust the whole of Martin's wide-ranging interests. When David A. Klarner edited a volume of mathematical recreations in tribute to Martin Gardner in 1981, he punningly gave it the title *The Mathematical Gardner*, but he hoped that it would later be accompanied by other volumes, such as *The Magical Gardner*, *The Literary Gardner*, *The Philosophical Gardner*, or *The Scientific Gardner*, in tribute to Martin Gardner's wide-ranging interests. He did not suggest *The Origami Gardner*, but it is a fact not often appreciated that paperfolding of a kind, though not strictly mainstream origami, was the origin of Martin's column in *Scientific American*.

Some time during 1956, Martin Gardner submitted to *Scientific American* a short article about "Hexaflexagons," curious folded paper devices, which had been discovered by chance by Arthur

H. Stone, an English post-graduate student at Princeton University in 1939, and which were only now beginning to become more widely known. Martin's article appeared in the December 1956 issue of *Scientific American*, and the editorial board of the journal was clearly impressed because Martin Gardner was asked for more articles. They followed without interruption from January 1957 until 1980. During 1981 he shared the column with Douglas Hofstadter, before retiring at the end of 1981. Hofstadter was the author of the very successful book Escher, Godel, Bach, which had appeared in 1979. Martin finally retired from his column at the end of 1981, apart from two further "surprise" columns in August and September 1983. The year 1956 also saw the publication of Robert Harbin's Paper Magic, and the exhibition of the paperfolding art of Akira Yoshizawa, arranged by Gershon Legman at the Stedelyjk Museum in Amsterdam, took place in 1955. Suddenly paperfolding was in the air and the new initiatives quickly led to the formation by Lillian Oppenheimer of the Origami Center, in New York in October 1958.

Martin Gardner was not primarily a paperfolder, any more than he was a mathematician or a scientist. His first enthusiasm was conjuring, and it was out of this that his other diverse recreational interests sprang, especially paperfolding.

Early Life

Martin Gardner was born in Tulsa, Oklahoma on October 21, 1914. His father, a geologist, introduced him to magic when he taught him the "Paddle Trick," which employs a table knife and several pieces of paper. Martin describes several versions in his *Encyclopedia of Impromptu Magic*. Before long he was inventing tricks of his own, and he was only 16, and still in high school, when he began contributing to the magic magazine, *The Sphinx*. His first article was "New Color Divination" in May 1930. Already, his contribution for the following August carried the somewhat precocious title "The Best Pocket Tricks of Martin Gardner." It included a version of the knife paddle move he had learned from his father.

While a boy, Martin enjoyed Frank Rigney's jokes in *American Boy's Magazine*. Frank Rigney was also a conjuror, but he is best known to paperfolders as the illustrator and coauthor with William D. Murray of *Fun with Paperfolding*, which was published in 1928.

For many years, it remained the best introduction to paperfolding in English and influenced many paperfolders who later became well-known, including Lillian Oppenheimer. Far from being merely the illustrator of *Fun with Paperfolding*, Frank Rigney was also the creator of several of the folds it included. Later, when Frank became the illustrator for *Hugard's Magic Monthly* for which Martin was contributing a regular column on magic, they became close personal friends. That was, however, in the future. Martin entered the University of Chicago where his principal studies were in philosophy. He was a resident of Hitchcock Hall and became a member of Phi Beta Kappa. He graduated with a Bachelor of Arts degree in 1936 and for a time continued as a post-graduate student.

In November 1935, while he was still an undergraduate, Martin Gardner wrote *Match-ic*, the first of his many publications. This was a slim booklet of tricks with matches. Match-ic contained nothing about paperfolding, but his next booklet, *After the Dessert*, which was first issued in a mimeographed version in 1940, and published in a printed version in 1941, contained a number of paper-related tricks. Not least of these tricks was the "Japanese Paper Bird," which Martin had found in Houdini's *Paper Magic* of 1922 and traced back to the supplement to the second edition of 1890 of Tissandier's Scientific Recreations, translated from the original French. There is a silly but baffling trick, whereby a dollar bill is mysteriously turned upside-down, a trick which Martin Gardner included in several subsequent publications. He also included a stunt with a dinner napkin that he frankly described as an "Improvised Brassiere." It was an old trick popular with conjurors and often politely bowdlerized and placed on the head as "Cat's Ears."

It appears that even at this age Martin Gardner envisaged a journalistic or literary career. He had already written a short story called "Thang" for his college literary magazine. After a short period of research, he took a job as a reporter for the *Tulsa Tribune* and also became a staff writer with the press relations department of the University of Chicago. However, war intervened in 1941. He joined the United States Navy and saw active service in a destroyer escort in the North Atlantic. (More recently, destroyer escorts have been reclassified as frigates.) A destroyer escort was a small ship that could act as a scout for the fleet, searching out enemy submarines. With the constant threat from Nazi U-boats and the frequent storm-force winds of the North Atlantic, it was far from being a comfortable job. In his later book, *Whys and Wherefores*, Martin gave a rare glimpse into his personal life when he wrote that his destroyer escort was a "ship small enough so that a sailor could really get to know the sea in a way quite different from that of the tourist who floats gently on the ocean in a huge hotel."

After the War

Following the War, Martin Gardner returned to Chicago to take up research again. But in 1946 he was aged 32 and scarcely a young post-graduate student. He resumed his writing and his short stories began to sell. Many appeared in *Esquire* magazine, and he also wrote for *Humpty Dumpty*, a magazine for children, for which he became contributing editor. In a short time he was able to devote himself to his freelance writing. In 1947 he moved to New York City, and for many years he lived in or near the city.

Martin Gardner married Charlotte Greenwald in 1952. They enjoyed 48 years together until she died in December 2000. They had two children: Jim, who became an assistant professor of educational psychology at the University of Oklahoma, and Tom, who became a freelance artist in Greenville, South Carolina.

Martin's book *New Mathematical Diversions*, which was published in 1966, contains a cryptic dedication to his wife:

> Evoly met
> To L.R. AHCROF
> emitero meno

In case anyone has tried unsuccessfully to translate it as Latin, the simple solution is that backwards it reads: "One more time for Charlotte my love." (It was the second book that Martin had dedicated to Charlotte, the first one being *Great Essays in Science*, a collection of classic scientific essays that he had edited.) Equally cryptic is the more straightforward dedication of his philosophical book of 1983, *The Whys of a Philosophical Scrivener*:

> Why do I dedicate this book to Charlotte?
> She knows.

For a time the Gardners lived at Hastings-on-Hudson in New York, in a street very appropriately called Euclid Avenue. In 1982 they moved south to the warmer climate of Hendersonville in western North Carolina. Martin Gardner and a new generation of grandchildren have been able to share in his countless tricks and illusions with mutual delight.

Conjuring

Martin Gardner's interest in paperfolding is rooted firmly in his earlier passion for conjuring. The association between the two is a common one and countless leading paperfolders have also been magicians. There is a sort of mind that is attracted to puzzles, illusions, mathematical structures, patterns, unexpected transformations, linkages, and paperfolding. Martin Gardner has such a mind, and intellectually it led him to wrestle with the awesome conundrums of philosophy. But he never lost touch with those trivial tricks and puzzles that lighten the burden of life.

One of the roots of modern paperfolding was in conjuring, and it can be traced back well into the nineteenth century. Indeed, "The Magic Fan" or "Troublewit," as it is usually known in English, is a kind of paperfolding that goes back much further than that. The "Japanese Paper Bird" (now invariably known as "The Flapping Bird"), which Martin Gardner included in *After the Dessert* was apparently introduced to the West by traveling Japanese conjurors, probably in the 1870s. There is still, however, some uncertainly about its true origin. In the 1920s, magicians like Will Blythe, Will Goldston, and Houdini introduced magic using paper or "paper magic" in their acts. Toward the end of the Second World War the folding of dollar bills exercised a fascination for conjurors, and instructions for tricks with dollar bills became a popular feature of the numerous magical magazines of the time. In his preface to Samuel Randlett's *The Best of Origami* (1963), Martin Gardner recalled attending a magic convention in the 1930s at which almost every magician present was wearing a finger ring with a large rectangular "jewel" that he had folded from a dollar bill.

Martin Gardner always sought the company of other magicians. While still at home in Tulsa, he had been a member of a group that included Logan Wait and Roger Montandon, and as soon as he moved to Chicago, he was able to attach himself to a very rich fraternity of magicians. He joined the Chicago Magic Table and at Christmas took a job in a department store demonstrating magic sets. Magic shops are always meeting places for conjurors, anxious to find the latest tricks, and Martin has told how he hung around Joe Berg's and Laurie Ireland's magic shops in Chicago. Laurie Ireland's shop was later to pass through his widow to Jay Marshall, a good friend of origami, who renamed it Magic Inc. The shop later became a meeting place for CHAOS, the Chicago Area Origami Society. Another place of meeting for conjurors was the Nanking Chinese Restaurant. During his time in Chicago, Martin's idol was Werner Dornfield and he later dedicated one of his books to "Dorny." Yet another feature of magical life in Chicago was the regular magical conventions. One wonders whether Martin had not chosen his university for the richness of its magical culture!

As soon as Martin moved to New York, he found a similar scene awaiting him. Lou Tannen's Magic Shop was one of the meeting places. Another was the apartment of Bruce and Bunny Elliott, who frequently played host to meetings of magicians. Bruce was the editor of *The Phoenix*, a magazine to which Martin Gardner became a frequent contributor. Indeed he was a frequent contributor to numerous magic magazines and inevitably became well-known to the conjurors of his generation. Only when other pressures built up did he relax his enthusiasm for practical magic, especially when his *Scientific American* column made increasingly voracious demands on him. It was, however, the move to North Carolina in 1982 that severely reduced his links with other magicians. Desirable though the move might have been for other reasons, there was, unfortunately, no community of conjurors in Hendersonville.

After his move to Hendersonville and the conclusion of his column for *Scientific American*, Martin Gardner took up again the fallacies perpetrated as science by self-appointed experts in many strange areas. His book *In the Name of Science* was first published in 1952 and in 1956 was republished as a paperback under the name *Fads and Fallacies in the Name of Science* by Dover Publications. In it he brought his rational thinking to bear to expose the fallacies of such ideas as "Pyramidology" and the "Flat-Earth Theory." He became a fellow of the Committee for the Scientific Investigation of Claims of the Paranormal and has written a regular column for its journal *The Skeptical Inquirer*.

Martin Gardner's line of practical magic was the uncomplex, requiring the minimum of preparation. He eschewed elaborate mechanical illusions, and concentrated on tricks that made use of everyday objects, which were exemplified in his *Encyclopedia of Im-*

promptu Magic. This "encyclopedia" originally appeared as a long series of short articles in *Hugard's Magic Monthly* and was later published as a book by Magic Inc. of Chicago in 1978. It probably gives an unbalanced view of Martin's conjuring, for it excludes card tricks and tricks that require practiced sleight of hand, but the book gives a good general view of his approach. Separate from this were Martin's many articles on magic that constantly appeared in several other magic magazines and that were later collected together and published in 1993 in *Martin Gardner Presents*, another large book of the same size as the *Encyclopedia of Impromptu Magic.*

Paperfolding

Paperfolding and paper tricks fitted in well with this scheme of things. Added to this, the basic geometry of folding paper fascinated Martin Gardner, the mathematician. His long series of contributions to *Hugard's Magic Monthly* began in 1948 and his contributions for February and September 1949 were on "Dollar Bill Folds" and "Stunts with Paper," respectively. For the "Dollar Bill Folds," Martin included a "Blowing Fish" that could be made to blow out a candle, a method of reducing the size of a bill by folding, and "the Mushroom," two folds in a dollar bill that convert George Washington into a mushroom. It was a popular trick of Martin, for he has included it in several of his books. The article on "Stunts with Paper" was less concerned with paperfolding as such. It included a circle of paper-cut bunnies that was a simple device for apparently changing the expression of a face drawn on paper, named "Movies." It also included a simple "Snapper," an animal's head made from a small piece of card, which has the strength to pick up quite large objects in its "jaws."

Frankly, none of these tricks is very good paperfolding, or "Origami" as it was later to be known through the founding of the Origami Center in New York in October 1958. Yet it was probably these articles that brought Martin Gardner to the attention of Gershon Legman. Legman was then still residing in New York before he moved to live in France in 1953. He had been avidly collecting information about paperfolding from all possible sources since 1945, and he wrote to anyone he thought might be able to tell him more about the subject or who might give him information for the list of books and papers about it that he was compiling. Gershon

Legman was not himself a magician, but he was acquainted with many people who were conjurors, including Cy Enfield, the film producer, with whom he had been at school. It was Cy Enfield who taught Legman the "Bow Tie" or "Lotus," which set him off on his intensive quest for paperfolding in 1945. (By a curious turn of events that amounted to a stroke of fate, it was Enfield who introduced Robert Harbin to Legman.) But if Legman discovered that Martin Gardner was interested in tricks and paperfolding, we may be sure that through their acquaintance, Martin was introduced to the wider knowledge of paperfolding being built up by Legman.

There was a marked increase in Martin's paperfolding activity. In 1952 he contributed three items, not to a magic magazine, but to the *Children's Digest*. These were "How to Make a Paper Boat," "Loop the Loop," and "You'll Get a Bang out of This!" I have not seen these but the first two were probably some sort of boat and a flying device and the third was the traditional banger. Then, in September 1952, Martin contributed a glider to *Parents' Magazine*. But this was not quite the traditional paper glider folded from paper, because it incorporated a modified nose that enabled it to be propelled by an elastic band. The refinement was, in fact, a discovery of Gershon Legman. Because *Children's Digest* and *Parents' Magazine* were not magic magazines, these articles were omitted from the collection of Martin Gardner's articles in *Martin Gardner Presents* and it appears that they have never been reprinted.

Gershon Legman published a preliminary edition of his bibliography in the magazine *Magicol* in May 1952, and his longer *Bibliography of Paper-folding* in a slender booklet later in the year. In the latter publication, he listed Martin Gardner's two contributions to *Hugard's Magic Monthly* in February and September 1949 and his three contributions to *Children's Digest* between January and May 1952. The contribution of Gershon Legman's modified glider by Martin Gardner to *Parents' Magazine* in September 1952 was pointedly not listed.

It would be wrong to imply that arising from these publications was an immediate intensification of Martin Gardner's interest in paperfolding. For him it remained, as it always had been, just an aspect of conjuring. We have noticed that some paperfolding was already included in Martin Gardner's early booklet *After the Dessert* (1940–1941). After the War, in 1949, he published another booklet in a similar vein called *Over the Coffee Cups*. This, too, contained one or two paperfolding items. One was the notorious folding of a dollar bill that brought together the words "GAL TENDER AND PRIVATE" reputed to be used by a traveler who was asking for company in his room. Another was a trick by Samuel Berland. Once more, this was scarcely paperfolding, but it employed a method of folding one dollar bill to look like two bills, creating an illusion whereby the second bill was made to vanish.

Another favorite paper gimmick of Martin Gardner's was the "Moebius Bands," which he often presented under the mysterious name of the "Afghan Bands." This name had been used by the famous Professor Hoffman as long ago as 1904, but its origin remained a mystery. A version of the "Afghan Bands" by Martin Gardner appeared in *Hugard's Magic Monthly* in December 1949. Moebius Bands are frequently made of paper, although they are not in essence paperfolding, and Gershon Legman did not think it appropriate to include the article in his bibliography. Rather, the "Afghan Bands" portend the way Martin Gardner's interest would develop in the future. He was to include them in articles in *Scientific American* and in later books written by him.

Hugard's Magic Monthly quickly became a focus for much of Martin Gardner's work. After contributing some 32 articles between July 1945 and February 1951, he began a monthly column that ran from March 1951 until March 1958. The articles reflected Martin's general and uncomplicated approach to conjuring. He named the series the "Encyclopedia of Impromptu Magic." I have not examined the individual copies of the magazine, but it seems that from the start they were envisaged as the basis for a comprehensive book of conjuring by the same name. As he came across them, Martin made a practice of jotting down notes about conjuring tricks on filing cards that he kept in shoe boxes. When he wished to write an article, he rapidly recovered the information from the filing cards. Martin originally hoped to revise the compilation in light of published books in order to give acknowledgments to inventors of ideas and to polish it up generally. But time (as ever) would not allow and the collected articles were eventually published by Magic Inc. of Chicago in 1978. For the most part the articles were reprinted "raw." There are a few revisions and added notes but on the whole they are very limited and uneven.

Several of the headings in the book version of the *Encyclopedia of Impromptu Magic* are relevant to paperfolding including those on "Bill Folds," "Handkerchief Folding," "Magazine," "Newspaper," and a very long section headed simply "Paper." This last heading falls into various subsections, such as paper magic, paper stunts, paper cutting, paperfolding, paper work, and geometrical curiosities. Much under this heading is related in some way to paperfolding in the wide sense of the term. Here, for instance, are a long section on the Moebius Bands, a short one on geometrical folding, and a section on hexaflexagons. Many of the items are very familiar and come from the international stock of traditional indoor recreations.

The section on "Paper Folding," as such, is eight pages long and begins with a brief introductory history that mentions Japan, Unamuno, and Froebel. There follows a short bibliography with tribute paid to Gershon Legman's longer bibliographies. It contains just 16 items ranging from Will Blyth's *Paper Magic* of 1920 to Samuel Randlett's *The Best of Origami* of 1963. Obviously the list of books had either been updated or compiled specially for the book form of the *Encyclopedia*. Nevertheless, it remains a very selective list.

Martin frankly describes the folded paper figures that he includes as only a selection, chosen because they can be animated in some amusing way. He freely acknowledges that figures of great realism and beauty may be found in the Oriental and Spanish works on paperfolding. (This statement is also dated in its curiously restricted reference to the Orient and Spain, and it has clearly not been updated to include reference to the Western creations in Samuel Randlett's *Art of Origami* (1961) and *Best of Origami* (1963) or Robert Harbin's *Secrets of Origami* (1963).

The models mentioned or reproduced include the bellows, the "hopping" frog, a boat that floats on water, a "Pop Gun," the "Paper Cup," the "Kettle" (in which water can be boiled), and the "Salt Cellar" in its various forms of "bug catcher" and "fortune-teller." They are all very interesting and Martin throws considerable light on each item. But what is significant is that it is merely a collection of existing folds, and apparently it does not include any folds of Martin's own devising.

An interesting feature of *Hugard's Magic Monthly*, and therefore of the *Encyclopedia of Impromptu Magic*, is that it was illustrated as a labor of love by Frank Rigney, previously mentioned as the coauthor and illustrator of Fun with Paper Folding by William D. Murray and Francis J. Rigney (1928). Martin pays a gracious tribute to him in his introduction to the book version of the Encyclopedia. It was an association that made them firm personal friends.

The installments of "The Encyclopedia of Impromptu Magic" ran in *Hugard's Magic Monthly* until March 1958, although Martin continued to contribute a few articles until September 1961. By March 1958, however, changes were in the air. Another magician, Robert Harbin, had published his book, *Paper Magic*, in England in 1956. This book, despite its title, was unambiguously about paperfolding and was not about paper magic in the sense of conjuring with paper. Harbin, too, had only recently become acquainted by correspondence with Gershon Legman. Then, in the summer of 1957, Mrs. Lillian Oppenheimer, who had been greatly impressed upon receiving a copy of *Paper Magic*, flew across the Atlantic to meet Robert Harbin in London, but she failed to meet Gershon Legman in France as she had hoped, because he was away from home. The world of Western paperfolding was suddenly set alight and in October of the following year, Mrs. Oppenheimer unexpectedly (perhaps not entirely unexpectedly) found herself the founder of the Origami Center of New York and the editor of a new journal called *The Origamian*.

The Origami Center

For such a spontaneous organization, the Origami Center was remarkably well-organized. By the second issue of the *Origamian* in November 1958, a long list of no less than 35 Honorary Members had been appointed. All of them were issued special printed membership cards. Among the list appears "Martin Gardner— Paperfolder, Author."

Clearly, for him to be chosen, Martin Gardner had made a significant impact on the young world of modern Western paperfolding. Up to 1958 his total achievements as an author about any subject could scarcely be described as prodigious, although his published books included *In the Name of Science* (debunking pseudoscientific notions), *Mathematics, Magic and Mystery* (on magical tricks using mathematics), and *Great Essays in Science* (a collection of classic essays of which he was merely the editor). He had not written a single book solely devoted to paperfolding. Lillian Oppenheimer was, perhaps, generous with her Honorary Memberships (she had a shrewd head for publicity), and we can infer that Martin Gardner had entered fully into that small circle that exchanged information about paperfolding during the months that preceded and followed the emergence of the Origami Center.

Martin Gardner was not the only magician to be chosen. Others were Robert Harbin, Paul Duke, and Jay Marshall of Magic Inc. Several other Honorary Members were not described as magicians, but, like so many paperfolders, they carried magic wands in their knapsacks, including Guiseppi Baggi, Shari Lewis, Robert Neale, and "Thok Sondergaard" of Denmark (who was no less than Thoki Yenn). Another Honorary Member was Lester Grimes of La Rochelle, New York, the doyen of paper magicians. Since before World War II, he had been billed as "The Paper Wizard," performing his act in a paper costume and using only paper equipment and materials. He was surely a friend of Martin Gardner's. Lester Grimes played an active part in the early days of the Origami Center.

There is no record that Martin Gardner attended any of the early meetings of the Origami Center, and he is not mentioned again in the five issues of the first volume of the *Origamian* that were published between November 1958 and March 1959. In fact, apart from occasional references to his books, he is seldom again mentioned in the many later issues of the *Origamian* after it resumed publication in the summer of 1961. There is, however, a report that Martin Gardner attended the Second Annual Origami Get-together at the Origami Center (meaning Lillian Oppenheimer's private apartment) on November 2–3, 1963, one of the earliest origami conventions to be held.

Another report in the *Origamian* for summer 1964 refers to Martin Gardner's short article on origami in *Encyclopaedia Britannica*, which first appeared in the edition published earlier in that year. The *Origamian* reports that Martin was considerably upset that he had written the article as long ago as 1959, but publication had been delayed. Although he had repeatedly asked to revise the article, the editors had adamantly refused to agree to this. As a result, the article was out of date before it appeared and it contained no mention of the distinguished American and other Western paperfolders who had emerged since 1959, not least of them Fred Rohm and Neal Elias. No doubt the editors of *Encyclopaedia Britannica* had their reasons, but Martin Gardner felt very embarrassed.

As it is, the article is interesting for a number of reasons. It reveals that in 1959 Martin Gardner was considered a sufficient authority on origami to be invited to write the article. The article mentions Akira Yoshizawa, Miguel Unamuno, and Vicente So-larzano Sagredo, the three most distinguished paperfolders before the formation of the Origami Center. Friedrich Froebel is mentioned in connection with the Kindergarten movement, and the Bauhaus is mentioned in connection with training students in commercial design. Arthur H. Stone is given credit for the discovery of flexagons. Two modern folders who are named are George Rhoads and Guiseppe Baggi. Altogether the article is a remarkably comprehensive, but compact, summary of paperfolding as it was in 1959. There is no mention of the Origami Center, but perhaps it was too soon for that. One wonders how Martin would have revised the article had he been allowed to do so in 1964, assuming that he would have been allocated no increase in space.

Not directly linked to the Origami Center, but certainly associated with it, was the exhibition "Plane Geometry and Fancy Figures" held at the Cooper Union Museum, New York from the beginning of June in the summer of 1959. The exhibition had already been planned when the summer of 1958 suddenly brought origami to the notice of the public in newspapers and on television. Because of this, Lillian Oppenheimer was invited to provide models for a section of the exhibition to be devoted to origami. She gathered models from the United States, Europe, and Japan. Martin Gardner was invited to contribute, but the catalogue lists only one model under his name, which was a flying bat. Although it was not apparent from the static exhibit, Martin has revealed that his bat had a secret. If its head was placed on one's fingertip, it balanced horizontally. The secret was a penny concealed in the tip of each wing. Of course, Martin may have submitted other models that were not selected for display by the Museum authorities. As befitted a public museum, they were rigorous in their selection of models they considered suitable for exhibition and even included only a fraction of the models by Yoshizawa that had been submitted to them. Nevertheless, it seems fair to say that Martin Gardner's very limited contribution to the exhibition is confirmation that notwithstanding his very creative contributions to card tricks and puzzles, he was not a creative paperfolder.

Scientific American

As the world of paperfolding was beginning to change in 1956 with the publication of Robert Harbin's *Paper Magic*, Martin Gardner's own world was also changing. At the time it did not seem significant—just one more magazine article among so many. Yet in retrospect it was the turning point in his life. Martin Gardner can have had no idea of the consequences of that first article that he wrote for *Scientific American* in December 1956, of the delight it would bring to millions of people worldwide, of the fame (and, let it be said, financial reward) it would bring, of the broadening of his own interests, or of the demanding challenge of writing a monthly column of such quality for 25 years. It is tempting to speculate about how many young people have been inspired by his column to take up mathematics as a career, or what advances in serious mathematics may have been stimulated by its regular disclosure of offbeat ideas in recreational mathematics.

The first of Martin Gardner's articles for *Scientific American* appeared in the issue for December 1956, with the title "Flexagons." It was about hexaflexagons in particular. An article about Moebius Bands appeared in June, 1957 and another about tetraflexagons in May, 1958. The article on "Origami" did not appear until July 1959. By then Martin had become so involved with his regular column in *Scientific American*, that he gave up his contributing editorship of *Humpty Dumpty*.

Hexaflexagons straddled the two realms of paperfolding and mathematics. They were discovered in 1939 by Arthur J. Stone, then a 23-year-old British research mathematician at Princeton University. He had trimmed the wider American file paper that he had bought to fit into his narrower British files. Then, he began to play with the excess strips of paper. After creasing them at 60-degree angles and interweaving them, he discovered that they formed flat hexagons that could be "flexed" to bring different faces of the paper successively into view. The Princeton mathematics department experienced a craze for what came to be named *flexagons* (obviously derived from the word *hexagons*) and Arthur Stone became the focus of a small group of fellow students who were fascinated by the mathematics of the new devices. Stone was joined by Bryant Tuckerman, John W. Tukey, and, not least, Richard P. Feynman, a genius who later achieved fame as a brilliant physicist. Together they analyzed the mathematics involved in flexagons and set out their theories in a comprehensive paper, which is said to have been a complete exposition of the subject. For some reason that has never been explained, this paper has never been published, and it has been left to others to publish analyses of their own. Martin Gardner's own account of hexaflexagons in his first article for *Scientific American* was not intended in any way to be comprehensive, but it is wonderfully informative and succinct.

Moebius Bands share an article with other curious topological models but the article has little to do with paperfolding. The bands are presented from the point of view of topology and as the basis of magical tricks.

Tetraflexagons are much less well-known than their cousins, the hexaflexagons, but Arthur Stone was interested in them, too. In another article, Martin Gardner points out that they have been known as a "double-action hinge" for centuries and toys based on the principle were marketed in the 1890s. He also mentions a tetraflexagon in the form of a puzzle that was copyrighted in 1946 by Roger Montandon of The Montandon Magic Company of Tulsa, Oklahoma. It was called "Cherchez la Femme," the puzzle being to find the picture of the young lady behind the facade of a grinning sailor. It was only in 1993, with the publication of the book, *Martin Gardner Presents*, that it was disclosed that the originator of this puzzle was Martin Gardner himself. Perhaps his reticence about the puzzle and its publication is explained by the fact that when eventually the lady is found, she is discovered to be *au naturel!*

The article on tetraflexagons also contains a full explanation and diagrams for a variant of the "Flexatube" puzzle in which a square tube of paper is turned inside-out by successive folding steps alone. It is revealed that this, too, was discovered by Arthur Stone while working on flexagons. No paperfolder fails to be fascinated by this magical folding device.

Martin Gardner's article on origami in *Scientific American* for July, 1959, gives a brief outline of the subject, describing it as "the ancient Japanese art of paper folding." In a few brief sentences, the article manages to mention Mrs. Oppenheimer, the Cooper Union Exhibition, the accomplishments of refined Japanese ladies, Lewis Carroll, and Miguel Unamuno, the Spanish philosopher, who wrote a mock-serious treatise on paperfolding. These are followed by the pentagonal knot in a strip of paper that conceals within it a mystical pentagram and by the far-from-simple-scientific problem of why, when we fold a sheet of paper, the crease is a straight line. Although less related to classic paperfolding, Martin also demonstrates how a parabola may be formed by successively folding one edge of a square of paper to a selected point that becomes the focus of a curve formed by the creases. Martin could not resist concluding his article with instructions for the "Flapping Bird." Written in 1956, the method was the old one of pre-creasing and crunching the points together, which was used by Tissandier and Houdini. Even though, at that time, Yoshizawa's scheme of different dotted lines to distinguish mountain and valley folds had not yet reached the West, nevertheless, Martin Gardner's diagrams are remarkably clear.

An article in *Scientific American* dated June 1960, which is entitled "Paperfolding and Papercutting," is mainly about dissections, but it also touches on papercutting or "kirigami," and it includes the famous dissection puzzle usually known as "Heaven and Hell."

After 1960, there was a long wait before Martin Gardner included anything more related to paperfolding in his *Scientific American* column. His column was changing, too. His earlier articles dealt with comparatively simple puzzles, tricks, and phenomena which, even if they might conceal hidden mathematical mysteries, were within the scope of understanding of any reasonably educated person. However, his later articles began to delve deeper, reflecting the growing appreciation that the playful exploration inspired by recreational mathematics could occasionally open up new vistas in advanced mathematics that were entirely unexpected and yet that sometimes unexpectedly proved to be of great value in newly emerging branches of science.

In April 1968, Martin reverted to play and wrote about "Puzzles and Tricks with a Dollar Bill." For these he went back to the beginning of his magical career, and included the trick of inverting a dollar bill and even the two folds that convert George Washington into a mushroom. There are also several mathematical tricks based on the serial number of a dollar bill. All of these are tricks that occur several times in different books of Martin Gardner's puzzles.

The following December, Martin turned again to another of his favorite subjects, Moebius Bands. His article reproduces two prints by the Dutch artist M. C. Escher and throws several new beams of light on an old subject, but there is little for paperfolders. Martin does, however, point out that a hexaflexagon is an interwoven Moebius band, something that is not immediately self-evident.

In May and September of 1971, Martin Gardner introduced two new paperfolding topics. The article for May 1971 is entitled "The Combinatorial Richness of Folding a Piece of Paper." It reveals the unexpectedly difficult mathematical problem of determining the number of ways in which a map, or for that matter a single strip of stamps, can be folded up. However, the article suddenly trans-forms itself into an account of the work of Robert Neale, a fellow magician and inventor of many paperfolding devices, most of which have an unusual "twist." Included are Robert Neale's "Beelzebug Puzzle," an ingenious puzzle based on a tetraflexagon, and the fa-mous "Sheep and Goats" paperfolding puzzle. Robert Neale's best-known trick, "Bunny Bill," is merely mentioned, but the address from which it can be obtained is given as Magic Inc. of Chicago.

"Plaiting Polyhedrons," which appeared in September 1971, out-lines the absorbing method of folding the Platonic solids from strips of paper. It is a subject that has been investigated from various an-gles by several paperfolders, and Martin Gardner's account whets the appetite. So far as is known, a comprehensive book on this far-from-negligible subject remains to be written.

One of the new mathematical topics that has emerged since World War II is that of fractals and one of Martin Gardner's last articles in *Scientific American* related to paperfolding is about the "Dragon Curve," which is a kind of fractal. Apparently this article was included in a series of "Nine logical and illogical problems to solve" in November 1967, but I have not seen the series. The part of the article about the "Dragon Curve" is reprinted in *Mathemati-cal Magic Show*, published ten years later in 1977. Martin Gardner demonstrates the method of creating the "Dragon Curve" by repeat-edly folding a piece of paper in half. There are, of course, physical limitations which, in practice, restrict this process to about seven folds, but the general theory of the "Dragon Curve" as a fractal is not invalidated.

Virtually all of Martin Gardner's articles in *Scientific American* have been reproduced in his volumes of scientific recreations. De-pending on which of Martin's books are included in the list, there are 15 or 16 collections of the *Scientific American* articles that were published over a period of 38 years by a variety of publishers in the United States and England. The first collection was *The Scien-tific American Book of Mathematical Recreations*, which appeared in 1959. In England it was published in 1961 with the title *Math-ematical Puzzles and Diversions from Scientific American*. The last volume in the series is *The Last Recreations* of 1997. Paperfolding is embedded in the books in just a few chapters. However, they demonstrate that just as Martin Gardner's own interests widened, so paperfolding has broadened its horizons, something that has been startlingly demonstrated by the recent explosion of interest in the mathematics of paperfolding in books and articles, in universities, and by the three international conferences devoted to the mathematics and science of paperfolding that have taken place so far in Italy, Japan, and California.

Some Other Interests

But Martin Gardner's interests have always spread far beyond conjuring, paperfolding, and mathematics. Sometimes his books on the most unlikely subjects have overtones of paperfolding. He has remained a philosopher all his life and, his book *The Whys of a Philosophical Scrivener* is an absorbing apology for his own personal philosophy. In it, he displays an unexpected appreciation of Miguel Unamuno, the great Spanish philosopher, poet and paperfolder, who died on New Year's Eve, 1936 to 1937, at the beginning of the Spanish Civil War. Martin is said to have been influenced in his ideas on theism by Unamuno, and we may wonder whether Martin and Unamuno shared a common way of thinking.

In the very different field of literary criticism, Martin Gardner annotated several popular classics, including the poems "The Ancient Mariner" and "The Night before Christmas." As might be expected, he was very attracted by the work of Lewis Carroll and made annotated editions of *Alice in Wonderland* and *Alice through the Looking Glass*, which have since been combined in a single volume. In them, Martin did not fail to refer to Lewis Carroll's own interest in paperfolding. However, he made sure that he took his information from Lewis Carroll's own diaries (which mention paper boats and bangers) and that he did not make Lewis Carroll into a "great and enthusiastic paperfolder" as some overenthusiastic commentators have done.

When we come to assess Martin Gardner's place in the history of paperfolding, we, too, must be careful not to exaggerate. He was not a creative folder, and he did not write a single book solely about paperfolding. Had he not achieved fame through his column in *Scientific American*, our perception of his part in the growth of Western origami is likely to have been much less.

Yet, Martin Gardner certainly did play an important part in the development of origami. In the 1930s and 1940s, he was one of the magicians who helped to build up the popularity of paperfolding stunts and tricks. He played a part in the swelling interest in paperfolding in the West after 1957 when Gershon Legman, Robert Harbin, and Lillian Oppenheimer linked up to form a firm international base for future development. Martin added a note of academic respectability to paperfolding through his article for *Encyclopaedia Britannica*, despite the undue delay in its publication.

Above all, it was Martin Gardner's handful of paperfolding articles in *Scientific American*, which were later reprinted with additions in his subsequent books, that brought paperfolding as a mixture of play, art, and mathematics to the notice of a new audience and that demonstrated once and for all that paperfolding was much more than a children's pastime.

Martin Gardner's supreme achievement was his ability to communicate difficult and often profound subjects with a few deft, but human, strokes of his pen. He removed fear from our approach to mathematics and science. We can be grateful that he grew up surrounded by the humble art of paperfolding and that he was able to show that it is not only fun to do, but that it, too, has its place in the greater world of mathematics and science and is not unworthy of our time and our interest.

Acknowledgments. This article first appeared in 1945 in the late and much lamented private magazine FOLD.

I express my deep gratitude to Martin Gardner, to whom I have submitted this article. He has graciously given it his approval and has suggested some minor additions and corrections that I have incorporated in the revised edition.

I also thank Mick Guy for reviewing this article and for suggesting corrections of some of the typographical errors of which I had inevitably fallen victim.

I, alone, remain responsible for the content and for all inaccuracies.

I shall welcome any corrections to this article and also any further information or anecdotes about Martin Gardner's involvement with paperfolding.

...Nothing but Confusion? Anamorphoses with Double Meaning

István Orosz

Anamorphosis is the Greek term for retransformation. In art history it is used for those works of art that were made distorted and unrecognizable through clever geometrical constructions. But when viewed from a certain point or through a reflecting object placed upon it, the hidden image appears in its true shape, that is, it goes through retransformation. Depending on which of the two ways of retransformation that is used, there are two main types of anamorphosis. The first is *perspective anamorphosis* (or *oblique anamorphosis*), which was already in use during the early Renaissance (the fifteenth century). The second is called *mirror anamorphosis* (or *catoptric anamorphosis*), which appeared at the same time as the mannerisms of the baroque era (the sixteenth century).

Hans Holbein's painting, *The French Ambassadors*, is perhaps the most famous example of anamorphosis. In it, a distorted shape lies diagonally across the bottom of the frame. Viewing this image from an acute angle transforms it into the plastic and almost three-dimensional image of a skull.

Holbein's many epigones followed the example of this 1533 painting, popularizing the oblique anamorphosis technique by the end of the sixteenth century. The anamorphoses of Nicholas Hillard,[1] the well-known miniature painter, are lost today. However, his former model, William Shakespeare, must have known them well. In *Richard the Second*, his reference to anamorphic distortion is so evident that we may be sure that the theater-goers in London were also familiar with the technique then referred to as "the perspective." (The terminus technicus "anamorphosis" was not used at that time. It was first mentioned in *Magia Universalis* by Gaspar Scott,[2] a German Jesuit who lived in the 1650s.)

> For sorrow's eye, glazed with blinding tears,
> Divides one thing entire to many objects;
> Like perspectives which, rightly gaz'd upon,
> Show nothing but confusion—eyed awry,
> Distinguish form![3]

I began to make experiments with anamorphoses quite some time ago: I drew the first one in the late 1970s. My interest was not only in resurrecting anamorphoses but also in improving and developing this old-fashioned genre.

"There is nothing but confusion" in Shakespeare's text means that in the picture, there is only the chaos of a confused, unrecognizable base image. Instead of having a confused image, my intention is to bring sense to the basic anamorphic picture, giving it meaning in itself, with its second reading revealed by viewing it from a different viewpoint—for example, looking at it through a special mirror. The ambiguous layers that appear with this approach make use of the connection or contrast between two images within the same picture that are independent of each other. This approach also brings a philosophical nature to these anamorphoses. In my etching of *The Theater of Shakespeare*, the two images complete each other in a thematic way. If we look at it straight (Figure 1, top), as we usually do in the case of a picture, we see a

---

[1]Nicholas Hilliard (1547–1619), court miniaturist, limnist, and engraver. The sitter in one of his portraits is thought to be Shakespeare.

[2]Gaspar Scott, German Jesuit physicist, student of Athanasius Kircher. His most interesting work is the *Magia universalis naturæet artis*, 4 vols., Würtzburg, 1657–1659, which contains a collection of mathematical problems and a large number of physical experiments, notably in optics and acoustics.

[3]William Shakespeare: *Richard II*, Act 2, Scene 2, lines 16–20.

Figure 1. *The Theater of Shakespeare.*

theater in the sixteenth century with actors, audience, and people looking around. On the basis of a historical picture,[4] the Swan Theater might have been like this. If we step to the right side of this exceptionally wide panoramic picture and look at it from an acute angle (Figure 1, bottom), from which the picture is seen as a narrow stripe, the elements of the theater not only disappear but are transformed into a portrayal of Shakespeare. The building becomes a portrait.

[4]The sketch was made by Johannes De Witt in about 1596.

Figure 2. *Jean de Dinteville.*

On the basis of Holbein's famous painting, I drew the French
ambassador, Jean de Dinteville. In my opinion Jean de Dinteville[5]
not only ordered the painting and modeled for it, but also had a
role in shaping the philosophical background of *The Ambassadors*.
Now let's look at this portrayal first from a low angle (Figure 2,
left), and then observe the way it becomes a still life while you turn
in front of it (Figure 2, right). We see—in a state of confusion—a
set of objects that are found in Holbein's painting or have some
references to them. Still, the thematic connection between the two
images is less exact than it was in the case of the Shakespeare-
theater.

---

[5]French ambassador representing his King, Francis I, in the court of Henry VIII
in London.

Figure 3. *Dürer in the Forest.*

This kind of method is unprecedented neither in my works nor in art history. Ever since an exhibition in Venice in 1987, it has been called the *Arcimboldo-effect*,[6] after the sixteenth-century painter from Milan, Giuseppe Arcimboldo.[7] You might call "an image within an image" the "picture of paranoia," a term suggested by Salvador Dali,[8] who played a great role in rediscovering Arcimboldo in the twentieth century. The early works of mine that employ this kind of technique are actually art-historical references to Dali, Arcimboldo, and Albrecht Dürer (who also drew hidden portraits within landscapes[9]). In addition there are references to M. C. Escher,[10] who was not a designer of ambiguous pictures but dealt with geometrical situations having multiple viewpoints. He is also an organic part of the above-mentioned list of names.

---

[6]The exhibition took place in Palazzo Grassi in the spring of 1987.

[7]Italian painter (b. ca. 1530, Milano, d. 1593, Milano).

[8]Salvador Dali (1904-1989), one of the best-known surrealistic painters, repeated the trick of Arcimboldo. He called it his "Paranoiac-Critical method," which is an interpretative disorder to reveal the double significance of things.

[9]For example, *View of Val d'Arco*, watercolor in c. 1495, and *Remains of a castle wall*, watercolor in 1514.

[10]Maurice Cornelis Escher (1898–1972), Dutch graphic artist and printmaker primarily known for his mathematical prints.

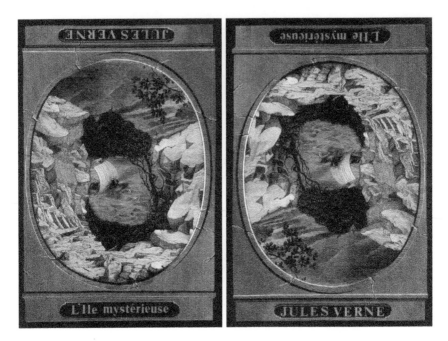

Figure 4. *Mysterious Island.* (See Color Plate I.)

The next few works of mine are not real anamorphoses: changes in the distance to the picture should not mean changes of the angle of your point of view when looking at them to reveal another meaningful layer. From close up, details receive the main emphasis (this time the etching *Dürer in the Forest*, shown in Figure 3, is purely a landscape), while from far off, you get a more complex impression of the whole picture than is revealed by the portrayal of the German master with Hungarian origins. Instead of moving either a step closer or further away, you can simply blink your eyes and the whole picture will then prevail over the details. The situation becomes a bit more complicated if it is not the viewer but the picture that changes its "point of view." In the case of the illustration *Mysterious Island*, there is a seashore with a sail pushed along by the wind (Figure 4, left). But if we are allowed to turn the image over, the portrait of Jules Verne, my favorite author in my childhood, will appear (Figure 4, right).

Speaking of Verne, there is another illustration made for his novel, *The Adventures of Captain Hatteras*, a polar landscape with

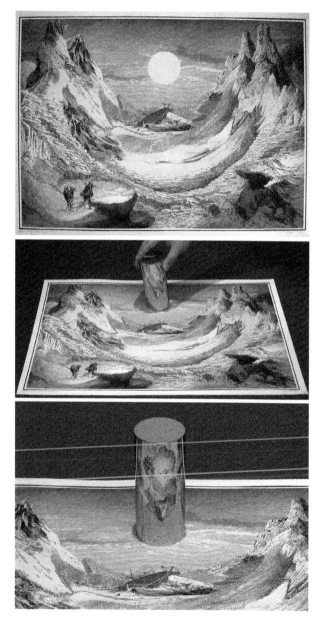

Figure 5. Anamorphosis illustration for *The Adventures of Captain Hatteras*.

snow at sunset (Figure 5, top). If you place a cylindrical mirror on the sun-disk in a horizontal position (Figure 5, middle), it will reflect the face of the French writer (Figure 5, bottom). This brings us to mirror anamorphoses. The best-known examples are those that employ cylindrical mirrors, but cone- or pyramid-shaped mirrors are also possible.

The images in Figure 6 of graphic construction show how the deformation takes place when only a mirror can show the original picture. The first book on anamorphoses, Le perspective curieuse (published in Paris, 1638) by Francois Niceron,[11] a friar monk, gives a detailed description of almost all the design strategies. I also follow his descriptions: somehow it is reassuring to know that your technique is the same as Leonardo's, Holbein's, Durer's, and their followers'.

M. C. Escher is portrayed in *The Well* in Figure 7. The original sketch was designed by Escher and his colleague, Bruno Ernst.[12] This sketch was of a mirror and a gate in a composition that allowed the place behind the gate to be revealed only in the mirror. Escher died before this idea was fully developed. The drafts were sent to me by Bruno Ernst to elaborate upon them and solve them for a graphic paper. I intended to have, instead of a one-layered image, a picture that also commemorated Escher. A mirror anamorphosis was designed in such a way that Escher's portrait was revealed in the mirror. Although he did not concern himself with anamorphoses but instead with mirrors and reflecting surfaces, he was famous for his "impossible objects": spatial formations that can be drawn without trouble but that are "nearly" impossible in space, or in a three-dimensional world. In an ordinary way of thinking, the concepts and constructions of Escher are beyond possibility, but for an "anamorphic" state of mind they are highly possible. The best-known impossible object is the *tribad*, or *Penrose triangle*, but the Swedish mathematician Oscar Reutersvärd may also come up in connection with the tribad.[13] Evidently, there is always a certain point, just as in the case of anamorphoses,

---

[11]Jean-Francois Niceron (1613–1646), French mathematician and painter, member of the Order of Minims.

[12]Also known as Hans de Rijk (1926–), Dutch mathematician and writer. He is the author of many books on M. C. Escher.

[13]The purest form of an impossible figure was drawn by Oscar Reutersvärd in 1934. This shape was the inspiration for an innumerable amount of impossible artworks, including those made by M. C. Escher. It is also accredited to Roger Penrose, because he published it first in 1958.

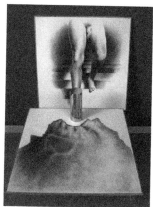
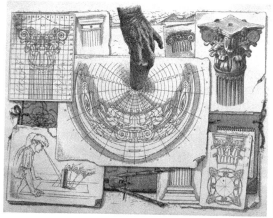

Figure 6. How to construct a deformation that shows the original image in a cylindrical mirror (left) and the final result (right): a foot (top) and a column (bottom).

where the tribad looks like a real and possible three- dimensional object. When drawing more complex "impossible objects," this is a special point of view that I always bear in mind.

Though our culture regards Escher as a graphic designer, he considered himself an amateur mathematician. Bruno Ernst, who is familiar with fine arts, also prefers to be regarded as a scientist concerned with optics and illusions. My picture in Figure 8 also presents a true physicist, although both the picture and the title

Figure 7. *The Well.*

may be misleading (*Self-portrait with Albert*). The chaos on my
desk is drawn from my personal point of view (and it is not at all
exaggerated), and I also appear in the round-shaped mirror. The
cylindrical mirror should be placed right on this mirror, that is,
on my face, to transform the distorted image of Albert Einstein
revealed in the mirror. Between the two layers of the picture, that
is, the horizontal and the vertical, there is no relationship unless
we think about such universal ideas as the contrast between order
and chaos, natural laws and the freedom of human beings.

Inspiration or design? When speaking about a work of art, you
may want to know which of these predominated in its creation. The
author of a work I chose to illustrate also asked this question, and
his answer was that a work of art can be created consciously "with
the precision and rigid consequence of a mathematical problem."
The writer was Edgar Allan Poe, and the work was *The Raven.* Poe
wrote an essay, "The Philosophy of Composition,"[14] in which he
offers a radical theory on the creative process as he describes what
lies behind his poem.

Someone viewing the illustration that I made for *The Raven* (see
Figure 9, left) will place a cylindrical mirror onto the point that
covers the bird's reflection in the wine glass; in so doing, they
emphasize the metaphoric interpretation of the poem and of the
picture. Edgar Allan Poe's virtual face is reflected in the mirror
(see Figure 9, right), made up of the objects lying horizontally, the

---

[14]The essay first appeared in the April 1846 issue of *Graham's Magazine.*

Figure 8. *Self-portrait with Albert.*

Figure 9. *The Raven.*

requisites of the illustration for *The Raven*. Once the cylinder is
raised, the face disappears, what is left are these scattered objects,
the shades, the man lying on his face and the empty room.

Poe claims in his essay that the most important effect to be
created in a work of art is that which allows it to be interpreted
backwards. His conclusion explains all the parts of the composition and their role in the whole. Poe was true to this in most
of his works. In fact, the same compositional scheme is at work

for an anamorphosis that has a second meaning, since by placing a cylindrical mirror onto the center of the paper, the viewer will realize why certain objects have been placed in the picture.

When designing my anamorphosis to Poe's poem, I attempted to work with a conscious and calculating mind, but I was also aware of the traps such childish logic might lead me into. All I could hope for was that the "inexplicable," too, always has and will have a role in all kind of creative work.

Previously, I mentioned utilizing the design methods of the seventeenth century's Niceron, even now, when superb computer programs are available (like Kent's Anamorph Me![15]) and distortion would be an easier and faster design method. Some years ago I worried that with Photoshop Polar Coordinates anybody could, within minutes, obtain an outcome similar to the works I spent days drawing. But actually the outcome is far from the same: the best computer programs are only able to solve the problem of distortion. A purely technical background cannot offer meanings or messages in the distorted image that are also connected with the original meaning of the base picture. "Nothing but confusion" is a task that only the art and soul of an artist can solve. At least for the time being . . . and hopefully for the impossibly enduring future . . . and a while afterward . . . .

---

[15]Anamorph Me! is a free, small software application developed by Phillip Kent, English mathematician, that can read images in the most common formats (e.g., JPEG, BMP) and carry out a range of anamorphic transformations on them— including oblique, cylindrical mirror, and conical mirror.

Part II

Ponder a Puzzle

Peg Solitaire with Diagonal Jumps

George I. Bell

An early fad of recreational mathematics occurred near the turn of the seventeenth century, when the one-person game "solitaire" swept through the French nobility in the court of Louis XIV. The evidence for this popular craze can be found in the art of the period. An engraving by Claude-Auguste Berey, dated 1697, depicts Princess Soubise posing by a board of the shape shown in Figure 1(a), and several other artists documented a similar scene [1]. These engravings are the earliest known references to the one-person game or puzzle now known as "peg solitaire."

We can't be sure exactly how the game was played in the seventeenth century, as no rules or solutions have survived from that time. Nonetheless, from later versions of peg solitaire, we can guess that the board was mostly lled with pegs, probably in a symmetrical pattern. The player then jumped one peg over another into an empty hole, removing the peg that was jumped over. The goal was to nish with as few pegs as possible, or in some predetermined pattern. Jumps must be made along rows or columns (not diagonally). We'll refer to this version without diagonal jumps as the *standard rules*.

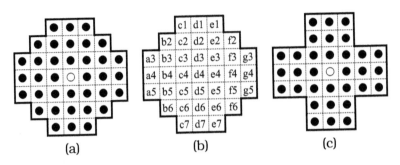

Figure 1. (a) The 37-hole board. (b) Hole coordinates. (c) The "standard" 33-hole board.

To denote a peg solitaire jump, we list the starting and ending coordinates (see Figure 1(b)) for the jump, separated by a dash. For example, the four jumps from the board position in Figure 1(a) under standard rules are d2-d4, b4-d4, f4-d4 or d6-d4.

One peg solitaire puzzle is the *central game*, which begins from a full board with one peg missing at the center (Figure 1(a)), with the goal of finishing with one peg in the center. Unfortunately, this puzzle cannot be solved to one peg on the 37-hole board under standard rules. This is not difficult to prove [1, 6], although it was probably not known in the seventeenth century. Eventually, it was realized that by modifying the board shape to Figure 1(c), the central game can be solved. This 33-hole board has become the standard board on which most people now play peg solitaire. This cross-shaped board is in some sense the smallest board with square-symmetry on which the central game is solvable [3].

Because the central game on the 37-hole board is unsolvable under standard rules, some have suggested that perhaps diagonal jumps were allowed back in the seventeenth century (for then the problem is solvable). However, if this did occur, it means that since the seventeenth century, the game has actually evolved away from diagonal jumps, and no such rule shift appears in the history of the game. It seems more likely that jumps have always been restricted to columns and rows.

Nonetheless, we'll now remove this restriction, and allow diagonal jumps in addition to standard jumps. We'll see that the game is a great deal more complicated, and loses a certain amount of elegance, but that interesting puzzles can still be found.

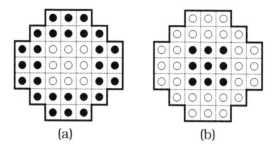

Figure 2. Puzzle 1: (a) Start. (b) Goal.

New Puzzles

Here we present a pair of puzzles based on the 37-hole board. From here on, diagonal jumps are allowed, and in fact required to solve the puzzles. These puzzles were found during a systematic investigation of peg solitaire allowing diagonal jumps [4]. To play these puzzles, you can purchase a 37-hole peg solitaire board (still commonly available), or play online [2].

Puzzle 1: The Big Central Game

Start from the board position in Figure 2(a) and play to finish at the board position in Figure 2(b).

At first glance, this puzzle may appear easy as you can jump pegs inward to fill the central region. However, near the finish a number of "extra" pegs remain that frustrate your task. It is easy to leave a peg stranded or to miss the goal configuration by one peg.

After attempting this puzzle, many people have the idea to try to play backwards from the final state. What is less obvious is that this is *exactly the same as the forward game!* In other words, if one has a solution to this puzzle, executing the jumps in the same direction, but in reverse order, is also a solution to the puzzle.

This *reversibility property* is also true of the central game on the 33-hole board under standard rules [8, pp. 122–135], and for the same reason. A peg solitaire jump can be thought of as reversing the state of, or *complementing*, three consecutive board locations. Suppose we begin from the board position that is the complement of the final board position (where every peg is replaced by a hole and vice versa)—for this puzzle the complement of the final board

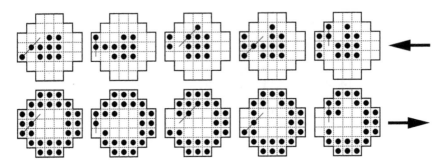

Figure 3. The end of a solution (top row, right to left). The start of the reversed solution (bottom row, right to left).

position happens to be the same as the starting board position. We can interpret the jumps played in reverse as normal jumps applied to the complement board position, so these jumps take one to the complement of the starting position (which is the final board position).

Figure 3 shows a sample reversed solution. The top row, reading right to left, shows the final five jumps of a solution to this puzzle. The bottom row shows the exact same jumps applied to the complement, or starting board position; these are the first five jumps of the reversed solution. Note that each board position in the top row, after the jump, is exactly the complement of the board position in the bottom row. (For all of this solution, see the end of this article.)

However interesting, these reversibility observations do not help to find a solution to the puzzle. More useful is the *Pagoda function* of Figure 4(a). This is a weighting of the board, carefully devised so that no jump increases its total. The Pagoda function P must satisfy the *Pagoda condition*

$$P(x) + P(y) \geq P(z)$$

for every possible solitaire jump from x over y into z. There are 168 different jumps possible on the 37-hole board (92 standard and 76 diagonal), and this condition must be satisfied for every one. You should check the Pagoda condition on a variety of jumps to convince yourself that it is satisfied. For example, for the jump $c1$-$c3$ we have $P(c1) = -1$, $P(c2) = 1$, and $P(c3) = 0$, and the Pagoda condition is satisfied.

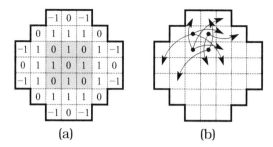

(a)   (b)

Figure 4. (a) A useful Pagoda function. (b) Forbidden jumps for Puzzle 1.

To calculate the value of this Pagoda function for a given board position, we simply sum the numbers where a peg is present. This total *can only decrease or stay the same as the game is played.* For Puzzle 1, the starting value of the Pagoda function is 4 (the sum of all values in the unshaded region), and the ending value is also 4 (the sum of all values in the shaded region). By definition, no jump can increase the value of the Pagoda function. However, we can also not afford a jump that decreases its value, for if we make such a jump we can never reach the final position. These *forbidden jumps* are shown in Figure 4(b)—any reflection or rotation of these jumps is also forbidden.

For example, if we begin with the jump c1-c3, then the peg at c3 can only be moved using the jump c3-e5, for any other jump starting from c3 loses something with regard to the Pagoda function. We can infer that any peg which jumps into the corners of the shaded goal area will almost certainly never jump again (it may, however, be removed by a jump). Alternatively, the starting jumps d1-d3, c2-e4 leave us in a board position with Pagoda total 3, and the final state is no longer reachable. This explains why the puzzle can seem difficult—after only two jumps, it is possible to reach a board position from which the puzzle can no longer be solved.

This Pagoda function also shows that the analogous Puzzle 1 on the standard 33-hole board can't be solved, even with diagonal jumps allowed. The central hole d4 starts out empty, and on this board the only way to fill it is using a nondiagonal jump that loses 2 with respect to the Pagoda function. On the 37-hole board, the same argument shows that the central hole d4 *must be filled by a diagonal jump.*

Puzzle 2: Nine-Man Solitaire

Start from the board position of Figure 2(b), and finish with one peg at any specified board location. At first we might think we need 37 separate solutions to solve this puzzle, one ending at each board location. Because of the symmetry of the board and the starting position, we need only find eight solutions, with all remaining cases being obtained by reflection and/or rotation. One possibility is to give eight solutions finishing at d1, d2, d3, d4, c1, c2, c3, and b2.

Versions of this puzzle can be traced back to Dudeney in 1917 [7], and the same puzzle can be found in several other puzzle collections [9, 10]. The starting configuration is the same, and the goal is to finish at the center d4. The problem is also stated on the 5×5 square board, but this is not an important difference.

Puzzle 2 tends to be somewhat easier than Puzzle 1 because there are only nine pegs at the start, and each solution has eight jumps. The Pagoda function of Figure 4(a) is not useful for this problem, because the starting value is 4, and the ending value is either -1, 0, or 1, depending on where we finish. Since the Pagoda function loses 3–5, this does not forbid any jumps. A useful technique is to identify the possible finishing pegs, and avoid jumping all of them. For example, if we want to finish at d4, the only peg capable of doing this is the one that begins at d4.

Combining the Puzzles

An interesting result comes from combining solutions to these two puzzles. Suppose we have a solution to Puzzle 2 finishing at d4; what happens when we play the jumps in reverse order? When reversed, we get not a solution to the same puzzle (as was the case for Puzzle 1), but a solution that takes us from the complement of the final state (a full board with one peg missing at d4, Figure 1(a)), to the complement of the initial state, which is the starting position of Puzzle 1 (Figure 2(a)).

Thus, by combining a reversed solution to Puzzle 2, a solution to Puzzle 1, and a second solution to Puzzle 2, we can go from a full board minus any peg and finish anywhere on the board. This gives us a constructive technique to solve any problem beginning with one peg missing and finishing with one peg anywhere on the board. By memorizing solutions, it is possible to solve any such problem by hand, although you have to be very good at reflecting, rotating, and reversing solutions to Puzzle 2 in your head.

This is quite different from peg solitaire under standard rules. For the standard 33-hole board, it is well known that if we begin from a full board with one peg missing at (x, y) (using Cartesian coordinates), then we can only finish at (x', y') where $x - x'$ and $y - y'$ are multiples of three. This is the so-called rule of three [6]. The combined result of our puzzles shows that when diagonal jumps are allowed, there is no longer any such restriction.

Short Solutions

After we have solved these two puzzles, we can try to find the "best" solution in some sense. Puzzle 1 begins with 28 pegs, and finishes with 9 pegs, so any solution must consist of exactly 19 jumps. However, when the same peg jumps *one or more* pegs, we call this one *move*. What is the solution to Puzzle 1 with the smallest number of moves?

It is not hard to show that a solution to Puzzle 1 must have at least 13 moves. First, note that the eight board locations c1, e1, a3, g3, a5, g5, c7, and e7 are special, because there is no way to jump over a peg at these locations. We call these eight locations the *corners* of the board. Also, let's call the central 3×3 finishing square the *target region*. To reach the final board position, the eight pegs that begin from the corners must be moved into the target region, using at least eight moves. Although they cannot occur consecutively, the net result of these eight moves can, at best, leave pegs at c3, e3, c5, and e5. This leaves five holes unfilled in the target region, and filling each requires a separate move, for a total of 13 moves.

This proves that a solution to Puzzle 1 must have at least 13 moves, and that any 13-move solution has the property that every move begins outside the target region and ends inside it. Even though we have some idea what it must look like, finding a 13-move solution is still difficult, but they do exist (see the end of this article).

A final question is: what is the solution to the central game with the smallest number of moves? This question is virtually impossible to answer by hand, as there are just too many complicated moves possible. Using a computer, we can find a solution in

13 moves. One can also prove analytically that no shorter solution is possible [4]. In contrast, for the 33-hole board under standard rules, an 18-move solution to the central game was discovered by Ernest Bergholt in 1912 [5], and was proved to be the shortest possible by John Beasley in 1964 [1, p. 135].

Summary

Over 300 years ago, the 37-hole board emerged during a puzzle fad in France. Since then it has been eclipsed by the 33-hole cross-shaped board. We have shown that the 37-hole board allows for some interesting puzzles when diagonal jumps are allowed. In some sense it is a more natural board to use when diagonal jumps are included.

Solutions

Puzzle 1

A 13-move solution is shown in Figure 5.

Here is an easily remembered solution found by hand (by John Beasley): c1-c3, d1-d3, e1-e3, f2-d4, g3-e3, g4-e4, g5-e5, d7-d5, f6-d6, e7-c5, c6-c4, a5-c5, c7-a5, b4-d2, b2-b4, a5-c3, d2-b4, a3-a5-c3 (Figure 3 shows the last five jumps of this solution). Note that the last six jumps are a "slanted six-purge," which is a standard peg solitaire six-purge [6] with diagonal rather than row jumps.

It is interesting to verify that reversing the jumps in either of the above solutions also gives a solution.

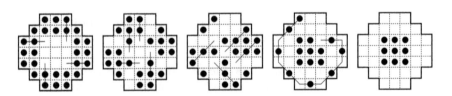

Figure 5. A 13 move solution to Puzzle 1.

Puzzle 2

All solutions have the minimum number of moves:

- To d1: d4-b4-d6-f4-d2, c3-e3, d5-f3-d3-d1;

- to d2: c4-e6, d4-f6-d6-b4-d2-d4-f4-d2;

- to d3: d4-d2-f4-d6-b4-d4, e4-c4, d5-b3-d3;

- to d4: d4-b4-d6-f4-d2, c3-e3, d5-f3-d3, d2-d4;

- to c1: d4-f4-d6, c4-c6-e6-c4-c2, e3-c3-c1;

- to c2: d4-b4-d6, e4-e6-c6-e4-e2-c4-c2;

- to c3: d4-b4-d2, e4-e2-c2-e4-e6-c4, c5-c3;

- to b2: e4-c2, d5-b3, d4-b2-d2-f4-d6-b4-b2.

Here is the solution to the central game in 13 moves: b2-d4, e1-c3, f4-d2, c1-e3, d6-f4-d2, g5-e5, g3-e1-c1-e3, b4-d6-f4-d2-b4, c7-c5, e7-e5, a5-c3, a3-a5-c7-e7-g5-g3-e3, d4-f2-d2-b2-b4-b6-d4-f4-d6-d4-d2-b4-d4.

Bibliography

[1] J. Beasley. *The Ins and Outs of Peg Solitaire*, paperback edition. Oxford, UK: Oxford University Press, 1992.

[2] G. Bell. "Diagonal Peg Solitaire for G4G7." http://www.geocities.com/gibell.geo/pegsolitaire/g4g7, 2006.

[3] G. Bell. "A Fresh Look at Peg Solitaire." *Mathematics Magazine* 80 (2007), 16–28.

[4] G. Bell. "Diagonal Peg Solitaire." *INTEGERS Electronic Journal of Combinatorial Number Theory* 7 (2007), G1. (Available at http://arxiv.org/abs/math.CO/0606122.)

[5] E. Bergholt. "English Solitaire." *The Queen* 131 (May 11, 1912), 666–667 and 807.

[6] E. Berlekamp, J. Conway, and R. Guy. "Purging Pegs Properly." In *Winning Ways for your Mathematical Plays*, Second Edition, Vol. 4, pp. 803–841. Wellesley, MA: A K Peters, 2004.

[7] H. Dudeney. *Amusements in Mathematics*, paperback edition. New York: Dover, 1958. (Originally published in 1917. See puzzle 229, "The Nine Almonds.")

[8] M. Gardner. *The Unexpected Hanging and Other Mathematical Diversions*. Chicago: Chicago University Press, 1991.

[9] D. Wells. *The Penguin Book of Curious and Interesting Puzzles*. New York: Penguin Books, 1992. (Reprinted by Dover in 2006.)

[10] N. Yoshigahara. *Puzzles 101: A Puzzlemaster's Challenge*. Wellesley, MA: A K Peters, 2004.

The Grand Time Sudoku and the Law of Leftovers

Bob Harris

I first saw jigsaw Sudokus as Mark Thompson's Latin Square Puzzles published in *GAMES World of Puzzles* in July 1996. Having seen no example larger than a 6 × 6, in 2001 I wrote a program to create a 7 × 7 for my Christmas card, and in 2005, as regular Sudoku returned to the U.S. from Japan, I began generating larger jigsaw variants, up to 9 × 9. This attracted the attention of one of the editors of this book, who suggested a jigsaw variant with MARTIN GARDNER as the clues. Figure 1 shows the best example I was able to find. The unique solution contains the letters GRAND TIME in each row, column, and region.

My goal was to find a puzzle that could be solved by simple logic, without resorting to any what-if conjectures. This puzzle *almost* achieves that goal. A single, quickly resolved conjecture is enough to solve it. The search produced this puzzle on the third day, after I made a lucky choice on the types of regions to allow. Another week and a half of searching produced no improvement. Several uniquely solvable puzzles were found, but none as easy as the one shown.

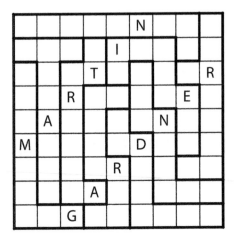

Figure 1. Jigsaw Sudoku variant with MARTIN GARDNER as the clues.

The next easiest puzzle found is shown in Figure 2, and contains a nice example of how to use my *law of leftovers* solving technique. The law, simply stated, is that wherever a group of regions overlaps some rows or columns, the parts outside the overlap—the leftovers—have to be the same.

Figure 2 shows the overlap between the first five columns and five regions. Five columns must contain five copies of each letter; the same is true of five regions. The overlap contributes the same letters to each collection, so the leftovers must complete each collection with the same letters. Thus we can infer that cells 85 and 95 contain an N, and since 85's region already contains N, the N must be at 95.

Figure 3 shows a second example of the law, with overlap between four columns and four regions. The two leftover regions must contain the same four letters, and these must be I, N, A, and G. For the column leftovers, the N we placed at 95 forces the N to 74, and the I must be at 94. For the region leftovers, 15 is the only possible spot for A in the fifth column, so the G must be at 17. Returning to the leftovers in Figure 2, we can infer that 85 must be G.

I've found the law to be very handy in solving low-clue jigsaw Sudokus, and it has picked up a little steam in Sudoku discussion groups now that jigsaws are becoming more popular. While every

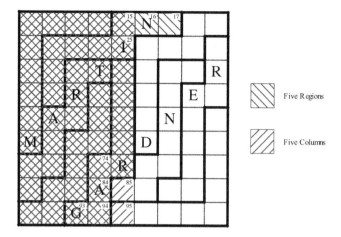

Figure 2. The overlap between the first five columns and five regions.

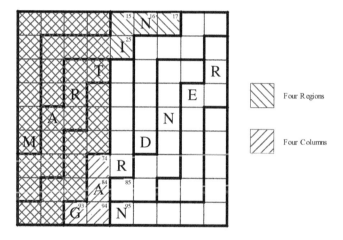

Figure 3. The overlap between the four columns and four regions.

relationship it reveals *could* be discovered by working through the logical consequences for each component row/column and region step by step, this simple geometric law provides a nice shortcut.

Patulous Pegboard Polygons

Derek Kisman, Richard Guy, and Alex Fink

A problem in a recent competition was: given a 2004 × 2004 square grid of dots, what is the largest number of edges of a convex polygon whose vertices are dots in the grid?

Of course, the question can be asked for any value of 2004, say n. For $n = 2$, 3, and 4, it's easy to see that the answers are $p = 2n$, as shown in Figure 1.

The next three values, shown in Figure 2, are not quite so obvious.

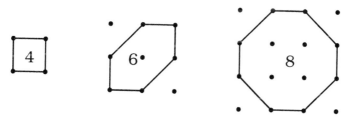

Figure 1. The best polygons for $n = 2$, 3, and 4.

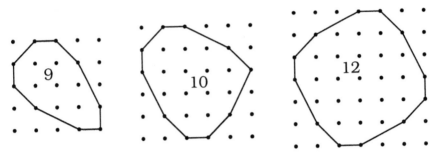

Figure 2. The best polygons for $n = 5$, 6, and 7.

How Can We Be Sure That They Are the Best?

As we go round the polygon, which we will always do counterclockwise, each edge has a certain *cost*, namely, the sum of its N-S and E-W components. The total cost must not exceed $4(n-1)$, i.e., at most $n-1$ in each of the four cardinal directions. There are just four available edges of cost 1, and four of cost 2. The former are used in all our polygons so far, except that the northbound one is missing when $n = 6$. Two of the latter are used when $n = 3$ and all of them subsequently, except for the northeasterly one when $n = 5$. Edges of cost 3 are like knight's moves in chess and occur two, three, and four times in our polygons for $n = 5$, 6 and 7, respectively. It's not too hard to see that whenever the answer is a multiple of 4 (as it is when $n = 2$, 4, or 7), we get it by using the cheapest possible edges. Edges have components a, b, say, so that their cost is $a + b = c$, which we keep to a minimum by assuming that $a \perp b$, that is, that a and b have no common factor bigger than 1. This is safe in the sense that if any polygon fits into a square grid then one with reduced sides will. We have seen that, when $c = 1$ or 2, there are just four different edge-directions. When $c \geq 3$, $a \neq b$ and there are just four orthogonal directions, $(\pm a, \pm b)$ and $(\pm b, \mp a)$, for each ordered pair (a, b).

The number of such pairs (a, b) for edges of cost $c = a + b$ is $\phi(c)$, Euler's *totient function*, the number of numbers from 1 to c that are prime to c. The first few values are the second row in the following table, which we have extended far enough to answer the opening question.

$c =$	1	2	3	4	5	6	7	8	9	10	11
$\phi(c) =$	1	1	2	2	4	2	6	4	6	4	10
$c\phi(c) =$	1	2	6	8	20	12	42	32	54	40	110
$\sum c\phi(c) =$	1	3	9	17	37	49	91	123	177	217	327
$n =$	2	4	10	18	38	50	92	124	178	218	328
$p = 4\sum\phi(c)$	4	8	16	24	40	48	72	88	112	128	168

$c =$	12	13	14	15	16	17	18	19	20	21
$\phi(c) =$	4	12	6	8	8	16	6	18	8	12
$c\phi(c) =$	48	156	84	120	128	272	108	342	160	252
$\sum c\phi(c) =$	375	531	615	735	863	1135	1243	1585	1745	1997
$n =$	376	532	616	736	864	1136	1244	1586	1746	1998
$p = 4\sum\phi(c)$	184	232	256	288	320	384	408	480	512	560

The third row is a quarter of the total cost of all edges of cost c, and the fourth row is the cumulative total, whose values are 1 less than values of n for which there is an optimal solution, shown in the fifth row.

Since we can introduce new edges in orthogonal sets of four, we can also include the values of n which, for each q, allow an optimal polygon with $p(n) = 4q$ edges:

$q =$	1	2	3	4	5	6	7	8	9	10	11	12
$n =$	2	4	7	10	14	18	23	28	33	38	44	50
$p =$	4	8	12	16	20	24	28	32	36	40	44	48

$q =$	13	14	15	16	17	18	19	20	21	22	23	24
$n =$	57	64	71	78	85	92	100	108	116	124	133	142
$p =$	52	56	60	64	68	72	76	80	84	88	92	96

$q =$	25	26	27	28	29	30	31	32	33	34	35	36
$n =$	151	160	169	178	188	198	208	218	229	240	251	262
$p =$	100	104	108	112	116	120	124	128	132	136	140	144

It is clear that such values of p are optimal. As $n = 1998$ corresponds to an optimal $p = 560$, we can answer the question from which we started. For $n = 2004$ an extra cost of $4(2004 - 1998) = 24$ is available, but we must use edges of cost > 21. The cost of one such edge cannot be shared between all four cardinal directions. One might try to replace the $(10, 11)$ vector with $(-6, 17)$ and $(16, -6)$, but the last is not primitive and is parallel to $(8, -3)$, so it doesn't give a new direction. But we can replace $(10, 11)$ and $(1, -20)$ with $(-1, -21)$, $(17, -5)$, and $(-5, 17)$, respectively increasing the N, S, E, and W components by $17 - 11$, $21 + 5 - 20$, $17 - 10 - 1$, and $5 + 1$—that is, 6 in each of the four directions—and increasing the cost by $3 \cdot 22 - 2 \cdot 21 = 24$, so that $p(2004) = p(1998) - 2 + 3 = 561$.

What If n and p Are Large?

For a picture, see Figure 5 at the end.

Theorem 330 of [3] tells us that when m is large,

$$\phi(1) + \phi(2) + \phi(3) + \cdots + \phi(m) \approx \frac{3m^2}{\pi^2},$$

so that

$$\phi(1) + 2\phi(2) + 3\phi(3) + \cdots + m\phi(m) \approx \frac{2m^3}{\pi^2},$$

and, when n is equal to the latter expression, p will be approximately four times the former. That is, $p \approx Cn^{2/3}$ with $C = 6 \cdot 2^{\frac{1}{3}}/\pi^{\frac{2}{3}} \approx 3.524206$. If we put $n = 1998$, we get $p = 559.05987$, and $n = 2004$ gives $p = 560.17855$, both correct if we round them up. It is rare for an asymptotic expression to give such a good result! Does the error get arbitrarily large?

p or n?

It may be more natural to invert the formula to

$$n \approx \frac{\pi p^{3/2}}{12\sqrt{3}}$$

and to invert the original question:

> What is the smallest $n \times n$ grid that will accommodate a convex p-gon?

If p is a multiple of 4, we know the answer.

In general, the total cost of the polygon must not exceed $4(n-1)$, so write

$$4(n-1) = t + e,$$

where t is the total cost of the p cheapest edges, and e is the extra expenditure that we must make. As e is nonnegative, we have the following lower bounds on n for given values of p:

$p =$	0	1	2	3	4	5	6	7	8	9	10	11	12	13	14	15
$t =$	0	1	2	3	4	6	8	10	12	15	18	21	24	27	30	33
$n \geq$	1	2	2	2	2	3	3	4	4	5	6	7	7	8	9	10

$p =$	16	17	18	19	20	21	22	23	24	25	26	27	28	29	30	31
$t =$	36	40	44	48	52	56	60	64	68	73	78	83	88	93	98	103
$n \geq$	10	_11_	_12_	_13_	14	_15_	_16_	_17_	18	20	21	22	23	25	26	27

$p =$	32	33	34	35	36	37	38	39	40	41	42	43	44	45	46	47
$t =$	108	113	118	123	128	133	138	143	148	154	160	166	172	178	184	190
$n \geq$	28	30	31	32	33	35	36	**37**	38	39	_41_	43	44	**46**	_48_	49

Figures 1 and 2 show that these minima are attained for $p \leq 16$. But, for $p = 17$, although the total cost of the p cheapest edges, $4 \cdot 1 + 4 \cdot 2 + 8 \cdot 3 + 4 = 40 = 4(11 - 1)$, is a multiple of 4, an 11×11 grid will not accommodate a 17-gon, because we would have to insert just one cost-4 vector into the optimal 16-gon and its cost has components 3 and 1, which can't be shared equally among the four cardinal directions. Similarly, in the cases $p = 18$, 19, 21, 22, 23, 42, and 46, the values of n (underlined in the table) have to be increased by 1 since the components of the available edges can't be distributed equally among the four cardinal directions. In fact, the 10-, 14-, and 18-grids required for $p = 15$, 19, and 23 will respectively accommodate 16-, 20-, and 24-gons.

Figures 3 and 4 show that the revised bounds for n can be attained for all $p \leq 48$ except for the bold entries in the last table under $p = 39$ and $p = 45$.

In Figure 3, $p = 13$ and 14 are clear. For $p = 17$ and 18 we have omitted the eight edges of costs 1 and 2; the labels "17 = 8 + 9" and "12 = 4 + 8" mean "eight edges omitted, nine edges shown" and "cost 4 omitted, cost 8 shown," respectively. These polygons have been obtained by adding edges to the optimal $p = 8$, $n = 4$ solution. For $p = 21$, 22, 25, and 26, respectively enlarge $p = 16$, $n = 10$ with the pentagon, hexagon, enneagon, and decagon from Figure 3. Notice that, in these last two cases, we could more economically have omitted the edges $(\pm 3, \pm 1)$ and $(\pm 1, \mp 3)$ from the diagrams, and obtain the solutions by adding the remaining pentagon and hexagon to an appropriate $p = 20$ solution, where, by "appropriate" we mean one that already contains these edges, that is, one that does *not* contain the edges $(\pm 1, \pm 3)$ and $(\pm 3, \mp 1)$. This same $p = 20$ solution can yield $p = 27$, by using the heptagon at the foot of Figure 3. For $p = 29$ and 30, enlarge $p = 24$ with the pentagon and hexagon.

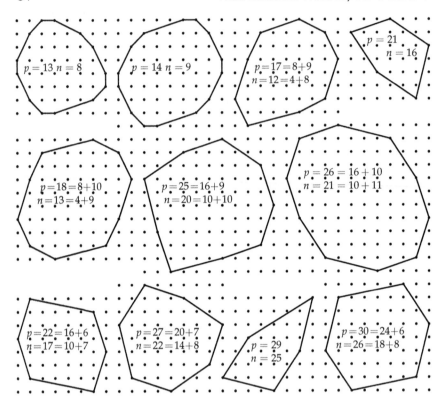

Figure 3. The smallest grids that accommodate p-gons, $13 \le p \le 29$.

In Figure 4, to obtain the solutions for $p = 31$ and 33, adjoin the heptagon and enneagon to the optimal $p = 24$ solution. Polygons $p = 34$ and $p = 35$ are obtained by adjoining the hexagon and heptagon to appropriate $p = 28$ solutions, that is, those containing $(\pm 4, \mp 1)$ and $(\pm 1, \mp 4)$ in the first case and $(\pm 4, \mp 1)$ and $(\pm 1, \pm 4)$ in the second.

For $p = 37$, replace the four dashed edges, $(3, -2)$, $(4, 1)$, $(1, 3)$, and $(-3, -1)$, in the optimal $p = 40$ solution with the edge $(5, 1)$. Polygons $p = 38$ and 41 are found by adjoining the hexagon and enneagon to appropriate $p = 32$ solutions; $p = 42$ from $p = 36$; and $p = 43, 45, 46$, and 47 from $p = 40$.

It remains to show that for $p = 39$ and $p = 45$ we can't attain the bounds in the table. A reduced 39-gon in a 37×37 grid would have $39 = 4 + 4 + 8 + 8 + c_5 + c_6 + c_7$ edges and total cost $4(37 - 1) =$

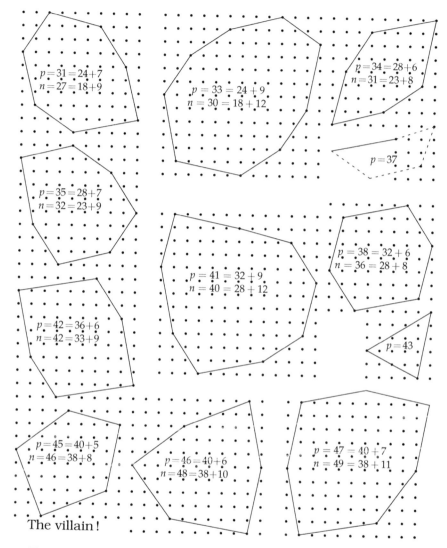

Figure 4. The smallest grids that accommodate p-gons, $31 \le p \le 47$.

$4 + 8 + 24 + 32 + 5c_5 + 6c_6 + 7c_7$, giving $(c_5, c_6, c_7) = (14, 1, 0)$. We must replace two cost-5 edges from the optimal $p = 40$ solution with a cost-6 edge. But the only ways in which we can do this, e.g., replace $(2, 3)$ and $(3, -2)$ with $(5, 1)$, yield a 39-gon in a 36×38 grid, with the correct cost, but unequally shared between E-W and N-S. In fact, $p = 39$ is the first example in the pattern of $e = 5$ cases where we are about to start or have just finished using edges of cost $4k + 2$. It's only $p = 45$ that is *the real villain*. A similar calculation for $p = 45$, $n = 46$ gives the solutions $(c_5, c_6, c_7, c_8) = (14, 7, 0, 0)$, $(15, 5, 1, 0)$, $(16, 3, 2, 0)$, and $(16, 4, 0, 1)$, none of which can be implemented on a 46×46 grid.

Plain Sailing

For $p > 48$ we shall always need to use edges of cost 7 or more, and, for $c \geq 7$, we always have $\phi(c) \geq 4$ and a good variety of slopes for our edges, and can always achieve the best bound we can hope for. Well, almost always! See $e = 6$, 5, and 4, below.

In fact, we've been spooked by the Law of Small Numbers [1, 2] and many of what have so far appeared to be exceptions are in fact part of a regular, though somewhat complicated, pattern. This is perhaps best expressed in the following summary. There's not room here for a complete proof, which tends to subdivide itself into an increasing number of cases. But the astute reader can reconstruct it from the examples that we've already seen, which are often early members of infinite sequences of polygons.

Summary

Recall that $4(n - 1) = t + e$; write $p = 4q + r$, where $r = \pm 1$, or $r = 2$, or $r = 0$; and write c for the cost of the most expensive of the p cheapest edges. Then,

- $e = 6$ just if $p = 45$.

- $e = 5$ just if r and c are both odd, but the cost of the $(p + 2)$th or the $(p - 1)$th cheapest edge is $4k + 2$ $(k > 0)$, as described for $p = 39$ above.

  The first few examples are $p = 39$, 49, 111, 129, 231, 257, 383, 409, ...

- $e = 4$ just if {r is odd and $c = 4k$} or {r is even, $c = 2k > 2$}, but the cost of one of the $(p - 2)$th or $(p + 3)$th cheapest edges is not c. That is, we have just started, or are about to finish, using even cost edges.

Examples that we've seen are $p = 17$, 18, 19, 21, 22, 23, 42, 46 and the next are $p = 73$, 74, 75, 77, 79, 81, 83, 85, 86, 87, 114, ...

- $e = 3$ just if r is odd and $c = 4k + r$.

  Examples are $p = 1$, 11, 15, 25, 29, 33, 37 and $p = 51$, 55, 59, 63, 67, 71, 89, ...

- $e = 2$ just if $\{r = 2$ and c is odd$\}$ or $\{r$ is odd and $c = 4k + 2\}$, except for the villain, $p = 45$.

  Examples are $p = 2$, 5, 7, 10, 14, 26, 30, 34, 38, 41, 43, 47 and $p = 50$, 54, 58, 62, 66, 70, 90, 94, 98, ...

- $e = 1$ just if r is odd, $c = 4k - r$, and e isn't already described above as being equal to 5.

  Examples are $p = 3$, 9, 13, 27, 31, 35, but not 39, not 49, and then 53, 57, 614, 65, 69, 91, 95, 99, 103, 107, not 111, not 129, but 133, 137, ...

- $e = 0$ just if $\{p = 6\}$ or $\{r = 0\}$ or if $\{r = 2, c$ is even, and e is not already described above as being equal to 4$\}$.

  Examples are $p = 4$, 6, 8, 12, 16, 20, ..., 68, 72, 76, 78, 80, 82, 84, 88, 92, ..., 108, 112, 116, 118, 120, 122, 124, 128, 132, ...

Reinventing the Wheel

Figure 5 shows what the polygons look like when n is large. As you go from one of the four cardinal points to one of the intermediate ones, the radius increases in the ratio $3\sqrt{2} : 4$, or by about 6%. The curve is very close in shape to that whose equation is $x^4 + y^4 + 3(x^2 + y^2) = 4$. Its actual equation is

$$\left\{(x^2 - y^2)^2 + 6(x^2 + y^2) - 7\right\}^2 = 32(x^2 + y^2 - 1)^3,$$

which is the best we can do at eliminating m from the parametric equations

$$x = \pm\frac{2m + 1}{(m + 1)^2}, \qquad y = \pm\frac{m(m + 2)}{(m + 1)^2}.$$

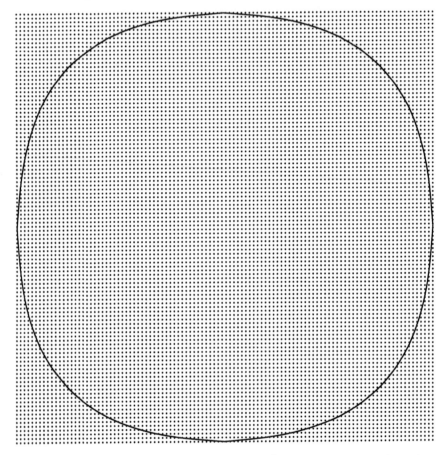

Figure 5. The 288-gon for $n = 736$—only 106^2 lattice points are shown!

Bibliography

[1] Richard K. Guy. "The Strong Law of Small Numbers." *Amer. Math. Monthly*, 95 (1988), 697–712.

[2] Richard K. Guy. "The Second Strong Law of Small Numbers." *Math. Mag.* 63 (1990), 3–20.

[3] G. H. Hardy and E. M. Wright. *An Introduction to the Theory of Numbers*, 4th edition. Oxford, UK: Clarendon Press, 1960.

Beamer Variant

Rodolfo Kurchan

In this article, we explore various types of *Beamer* puzzles, which involve drawing beams in grids in which some of the cells are the origins of the beams (called *capsules*).

Beamer

Each numbered capsule sends forth one or more beams in horizontal and vertical directions. Numbers indicate how many squares are touched by the corresponding capsule's beams. Squares containing capsules are not counted. Beams do not cross capsules and do not overlap or intersect each other. Each empty square is touched by exactly one beam.

Example:

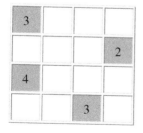
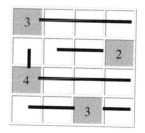

See Figures 1 and 2 for Beamer problems to try yourself. (Solutions provided at end of article.)

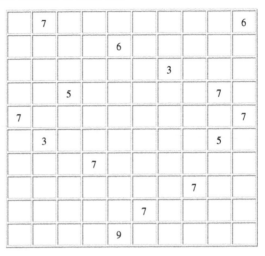

Figure 1. Problem 1.

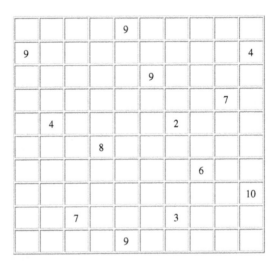

Figure 2. Problem 2.

Distraction Beamer

All rules of beams are the same, except that in this variant you find two numbers on each capsule. Only one of each pair of numbers indicates how many squares are touched by the capsule's beams—the other number is distraction.

Example:

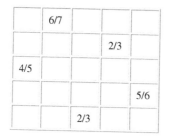 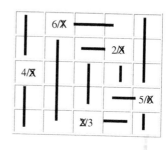

Figures 3 and 4 provide two Distraction Beamer problems.

Figure 3. Problem 3.

								8/16	
	5/10								
5/10									
				5/10					
					4/7				
			5/10						
		5/10							
						4/7			
5/10									
								10/15	

Figure 4. Problem 4.

Beamer Under Cover

One or more horizontal or vertical lines are drawn from each lettered capsule. Each letter represents a number (different letters stand for different numbers), and this number indicates how many squares are touched by its lines; squares with capsules are not counted. Lines do not cross capsules and do not overlap or intersect each other. Each empty square is touched by exactly one line.

Example:

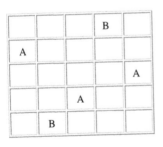
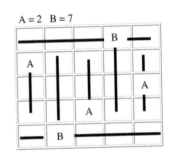

Figures 5 and 6 provide two Beamer Under Cover problems.

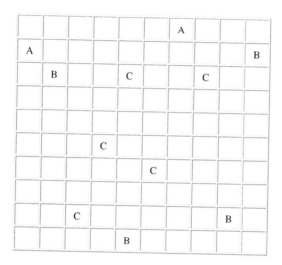

Figure 5. Problem 5.

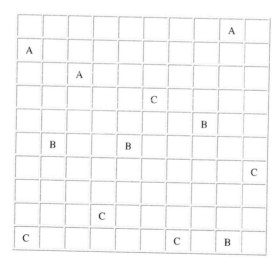

Figure 6. Problem 6.

Battleship Beamer

The beam rules are as in the previous problems; in this variation
the numbers outside the grid indicate the sum of the numbers that
are in the capsules in each row or column.

Example:

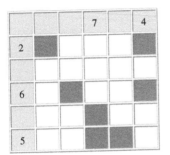 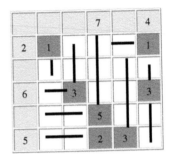

Figures 7 and 8 provide two Battleship Beamer problems.

Figure 7. Problem 7.

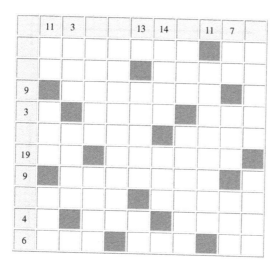

Figure 8. Problem 8.

Solutions

The solutions to Problems 1–8 are shown in Figures 9–16, respectively.

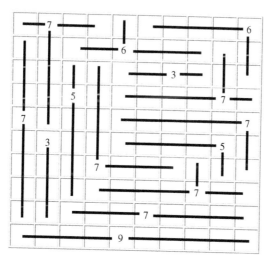

Figure 9. Solution to Problem 1.

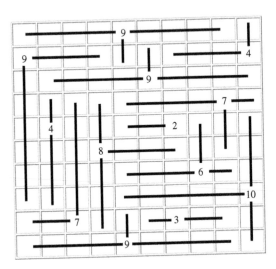

Figure 10. Solution to Problem 2.

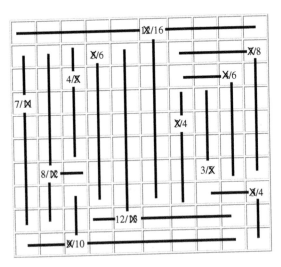

Figure 11. Solution to Problem 3.

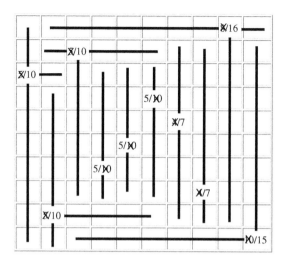

Figure 12. Solution to Problem 4.

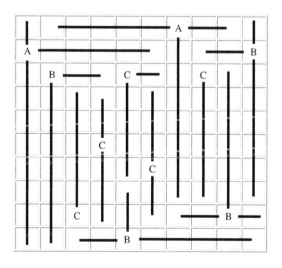

Figure 13. Solution to Problem 5: A = 14, B = 9, C = 5.

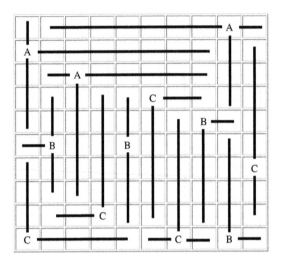

Figure 14. Solution to Problem 6: A = 11, B = 5, C = 7.

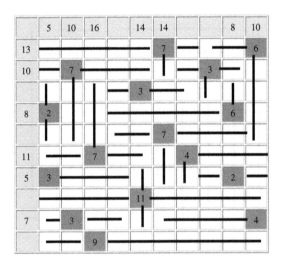

Figure 15. Solution to Problem 7.

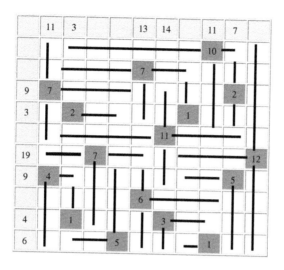

Figure 16. Solution to Problem 8.

Packing Equal Circles in a Square

Péter Gábor Szabó

The problem of densest packing of equal circles in a square is the following: locate $n \geq 2$ equal and nonoverlapping circles in a square in such a way that the radius of the circles is maximal. In other words, place n points in a square, such that the minimum of the pairwise distances is maximal. It is easy to understand what the problem is, but with an increasing number of circles the solution is very difficult. The proven optimal packings are known up to $n = 30$ circles, using artful computer-aided optimization methods [4].

Here we will focus on three special cases of this packing problem that have a bearing on the number 7. The circle packings will be investigated in the unit square.

Give a Proof of Optimality for Seven Circles!

Based on mathematical tools (without computer) there are known proofs of optimality of the packings only for $n = 6$, 8, 9, 14, and 16 circles (up to 5 circles the problem is trivial). The first hard

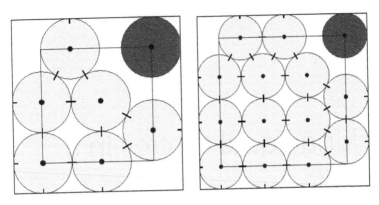

Figure 1. The densest packing of 7 and 14 equal circles in a square.

case is when the number of circles is 7 (Figure 1). J. Schaer had already given a proof in a letter to Leo Moser in 1964, but he never published it, because he thought it was an "ugly proof." Later, K. J. Nurmela and P. R. J. Östergård solved it by computer in 1996 [2]. It is interesting that this optimal packing contains a free circle (or rattle), and the structure of the packing is similar to the optimal packing for 14 circles.

Probably many people believe that if the number of circles in the square is increasing, the circle radius in the optimal circle packing is decreasing. Perhaps this is true, but nobody has proved it yet. A strange phenomenon occurs in the analogous problem of finding the densest circle packings in a circle. Here the circle radii have the same value in the optimal packings for both $n = 6$ and $n = 7$ circles (Figure 2).

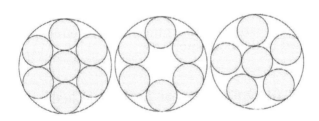

Figure 2. Optimal packings of seven and six circles in a circle with the same radii.

What Is the Exact Radius for 7 + 7 + 7 = 21 Circles?

Sometimes it is not easy to find the exact radius (not just an approximate value) for an optimal packing, in spite of knowing the structure of the packing. For up to $n = 20$ circles, here are some exact optimal values:

$$r_2 = \frac{2 - \sqrt{2}}{2}$$

$$r_3 = \frac{8 - 5\sqrt{2} + 4\sqrt{3} - 3\sqrt{6}}{2}$$

$$r_4 = \frac{1}{4}$$

$$r_5 = \frac{\sqrt{2} - 1}{2}$$

$$r_6 = \frac{6\sqrt{13} - 13}{46}$$

$$r_7 = \frac{4 - \sqrt{3}}{13}$$

$$r_8 = \frac{1 + \sqrt{2} - \sqrt{3}}{4}$$

$$r_9 = \frac{1}{6}$$

$$r_{11} = \frac{176 - 9\sqrt{2} - 14\sqrt{3} - 13\sqrt{6} - 2\sqrt{-16523 + 12545\sqrt{2} - 9919\sqrt{3} + 6587\sqrt{6}}}{568}$$

$$r_{12} = \frac{15\sqrt{34} - 34}{382}$$

$$r_{14} = \frac{6 - \sqrt{3}}{33}$$

$$r_{15} = \frac{8 - 5\sqrt{2} + 4\sqrt{3} - 3\sqrt{6}}{4}$$

$$r_{16} = \frac{1}{8}$$

$$r_{18} = \frac{12\sqrt{13} - 13}{262}$$

$$r_{20} = \frac{65 - 8\sqrt{2}}{482}$$

But what about $n = 10, 13, 17$, and 19 circles?

In these cases there are known minimal polynomials (polynomials of minimal degree), where the first positive root is the desired radius [3]. For example, for $n = 19$ circles the minimal polynomial is

$$P_{19}(r)$$

$$= 13694976r^{10} - 36781056r^9 + 50776320r^8 - 47372448r^7 + 30864436r^6$$
$$- 13692280r^5 + 4031385r^4 - 766866r^3 + 90131r^2 - 5954r + 169.$$

Minimal polynomials are known for some other cases ($n = 10$, 13, and 17) but are unknown for $n = 21$ circles (Figure 3) in spite of our knowing the structure of the optimal solution.

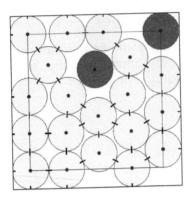

Figure 3. The densest packing of 21 circles in a square.

The 7 × 7 Circles Solution Is Not a Grid!

If you compare the structures of the optimal and the best-known circle packings [1], you can recognize repeated patterns. Perhaps the most conspicuous pattern belongs to the square numbers $n = 4, 9, 16, 25,$ and 36, where the circles have a $k \times k$ $(n = k^2)$ quadratic structure (Figure 4). In these packings the radius of the circles is

$$r_{k^2} = \frac{1}{2k} \quad (2 \leq k \leq 6).$$

It is interesting that the density of these circle packings is always equal to a constant, because the sum of the areas of the circles is

$$k^2 \left(\frac{1}{2k} \right)^2 \pi = \frac{\pi}{4}.$$

A surprising result is that the best-known circle packing for $7 \times 7 = 49$ circles is not a grid packing. If we used the grid pattern of Figure 4, the radius would be

$$\frac{1}{2} \approx 0.07142857,$$

but in the packing of Figure 5, the radius is greater than 0.07169268, and this is greater than 1/14.

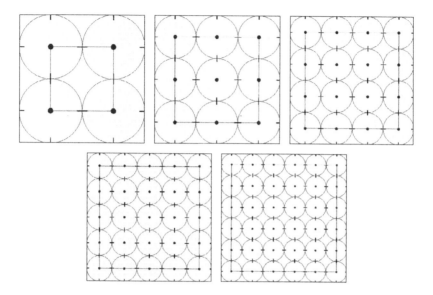

Figure 4. The densest packing of equal circles in a square for 4, 9, 16, and 25, and the best-known packing for 36 circles.

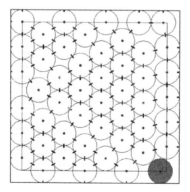

Figure 5. The best-known packing for 49 circles.

Bibliography

[1] R. L. Graham and B. D. Lubachevsky. "Repeated Patterns of Dense Packings of Equal Circles in a Square." *The Electronic Journal of Combinatorics* 3:16 (1996), 211–227.

[2] K. J. Nurmela and P. R. J. Östergård. "More Optimal Packings of Equal Circles in a Square." *Discrete & Computational Geometry* 22 (1999), 439–457.

[3] R. Peikert, D. Würtz, M. Monagan, and C. de Groot. "Packing Circles in a Square: A Review and New Results." In *System Modelling and Optimization*, Lecture Notes in Control and Information Sciences 180, edited by P. Kall, pp. 45–54. Berlin: Springer-Verlag, 1992.

[4] P. G. Szabó, M.Cs. Markót, T. Csendes, E. Specht, L. G. Casado, and I. García. *New Approaches to Circle Packing in a Square: With Program Codes*, Springer Optimization and Its Applications 6. New York: Springer, 2007.

Part III

Bring a Friend

Uncountable Sets and an Infinite Real Number Game

Matthew H. Baker

We give a short proof of the well-known fact that the unit interval $[0,1]$ in $\mathbb{R}$ is uncountable by means of a simple infinite game. We also show using this game that a (nonempty) perfect subset of $[0,1]$ must be uncountable.

The Game

Alice and Bob decide to play the following infinite game on the real number line. A subset S of the unit interval $[0,1]$ is fixed, and then Alice and Bob alternate playing real numbers. Alice moves first, choosing any real number a_1 strictly between 0 and 1. Bob then chooses any real number b_1 strictly between a_1 and 1. On each subsequent turn, the players must choose a point strictly between the previous two choices. Equivalently, if we let $a_0 = 0$ and $b_0 = 1$, then in round $n \geq 1$, Alice chooses a real number a_n with $a_{n-1} < a_n < b_{n-1}$, and then Bob chooses a real number b_n with $a_n < b_n < b_{n-1}$. Since a monotonically increasing sequence of real numbers that is bounded above has a limit (see [8, Theorem 3.14]),

$\alpha = \lim_{n \to \infty} a_n$ is a well-defined real number between 0 and 1. Alice wins the game if $\alpha \in S$, and Bob wins if $\alpha \notin S$.

Countable and Uncountable Sets

A nonempty set X is called *countable* if it is possible to list the elements of X in a (possibly repeating) infinite sequence $x_1, x_2, x_3, \ldots$. Equivalently, X is countable if there is a surjective function from the set $\{1, 2, 3, \ldots\}$ of natural numbers onto X. The empty set is also deemed to be countable. For example, every finite set is countable, and the set of natural numbers is countable. A set that is not countable is called *uncountable*. Cantor proved, using his famous *diagonalization argument*, that the real interval $[0, 1]$ is uncountable. We will give a different proof of this fact based on Alice and Bob's game.

Proposition 1. *If S is countable, then Bob has a winning strategy.*

Proof: The conclusion is immediate if $S = \emptyset$. Otherwise, since S is countable, one can enumerate the elements of S as $s_1, s_2, s_3, \ldots$. Consider the following strategy for Bob. On move $n \geq 1$, he chooses $b_n = s_n$ if this is a legal move, and otherwise he randomly chooses any allowable number for b_n. For each n, either $s_n \leq a_n$ or $s_n \geq b_n$. Since $a_n < \alpha < b_n$ for all n, we conclude that $\alpha \notin S$. This means that Bob always wins with this strategy! □

If $S = [0, 1]$, then clearly Alice wins no matter what either player does. Therefore we deduce the following.

Corollary 1. *The interval $[0, 1] \subset \mathbb{R}$ is uncountable.*

This argument is in many ways much simpler than Cantor's original proof!

Perfect Sets

We now prove a generalization of the fact that $[0, 1]$ is uncountable. This will also follow from an analysis of our game, but the analysis is somewhat more complicated. Given a subset X of $[0, 1]$, we make the following definitions:

- A *limit point* of X is a point $x \in [0,1]$ such that for every $\epsilon > 0$, the open interval $(x - \epsilon, x + \epsilon)$ contains an element of X other than x.

- X is *perfect* if it is nonempty[1] and equal to its set of limit points.

For example, the famous middle-third *Cantor set* is perfect (see [8, §2.44]). If $L(X)$ denotes the set of limit points of X, then a nonempty set X is closed $\Leftrightarrow L(X) \subseteq X$, and is perfect $\Leftrightarrow L(X) = X$. It is a well-known fact that every perfect set is uncountable (see [8, Theorem 2.43]). Using our infinite game, we will give a different proof of this fact. We recall the following basic property of the interval $[0,1]$:

($\star$) Every nonempty subset $X \subseteq [0,1]$ has an *infimum* (or *greatest lower bound*), meaning that there exists a real number $\gamma \in [0,1]$ such that $\gamma \leq x$ for every $x \in X$, and if $\gamma' \in [0,1]$ is any real number with $\gamma' \leq x$ for every $x \in X$, then $\gamma' \leq \gamma$. The infimum γ of X is denoted by $\gamma = \inf(X)$.

Let's say that a point $x \in [0,1]$ is *approachable from the right*, denoted $x \in X^+$, if for every $\epsilon > 0$, the open interval $(x, x + \epsilon)$ contains an element of X. We can define *approachable from the left* (written $x \in X^-$) similarly using the interval $(x - \epsilon, x)$. It is easy to see that $L(X) = X^+ \cup X^-$, so that a nonempty set X is perfect $\Leftrightarrow X = X^+ \cup X^-$.

The following two lemmas tell us about approachability in perfect sets.

Lemma 1. *If S is perfect, then* $\inf(S) \in S^+$.

Proof: The definition of the infimum in ($\star$) implies that $\inf(S)$ cannot be approachable from the left, so being a limit point of S, it must be approachable from the right. □

Lemma 2. *If S is perfect and* $a \in S^+$, *then for every* $\epsilon > 0$, *the open interval* $(a, a + \epsilon)$ *also contains an element of* S^+.

Proof: Since $a \in S^+$, we can choose three points $x, y, z \in S$ with $a < x < y < z < a + \epsilon$. Since $(x, z) \cap S$ contains y, the real number

---

[1]Some authors consider the empty set to be perfect.

$\gamma = \inf((x, z) \cap S)$ satisfies $x \leq \gamma \leq y$. If $\gamma = x$, then by $(\star)$ we have $\gamma \in S^+$. If $\gamma > x$, then $(\star)$ implies that $\gamma \in L(X)$ and $(x, \gamma) \cap S = \emptyset$, so that $\gamma \notin S^-$ and therefore $\gamma \in S^+$. $\square$

From these lemmas, we deduce:

Proposition 2. *If S is perfect, then Alice has a winning strategy.*

Proof: Alice's only constraint on her nth move is that $a_{n-1} < a_n < b_{n-1}$. By induction, it follows from Lemmas 1 and 2 that Alice can always choose a_n to be an element of $S^+ \subseteq S$. Since S is closed, $\alpha = \lim a_n \in S$, so Alice wins! $\square$

From Propositions 1 and 2, we deduce:

Corollary 2. *Every perfect set is uncountable.*

Further Analysis of the Game

We know from Proposition 1 that Bob has a winning strategy if S is countable, and it follows from Proposition 2 that Alice has a winning strategy if S contains a perfect set. (Alice just chooses all of her numbers from the perfect subset.) What can one say in general? A well-known result from set theory [1, §6.2, Exercise 5] says that every uncountable *Borel set*[2] contains a perfect subset. Thus, we have completely analyzed the game when S is a Borel set: Alice wins if S is uncountable, and Bob wins if S is countable. However, there do exist non-Borel uncountable subsets of $[0, 1]$ that do not contain a perfect subset [1, Theorem 6.3.7]. So we leave the reader with the following question (to which the author does not know the answer):

Question. Do there exist uncountable subsets of $[0, 1]$ for which: (a) Alice does not have a winning strategy; (b) Bob has a winning strategy; or (c) neither Alice nor Bob has a winning strategy?

---

[2]A Borel set is, roughly speaking, any subset of $[0, 1]$ that can be constructed by taking countably many unions, intersections, and complements of open intervals; see [8, §11.11] for a formal definition.

Related Games

Our infinite game is a slight variant of the one proposed by Jerrold Grossman and Barry Turett in [2] (see also [6]). Propositions 1 and 2 above were motivated by parts (a) and (c), respectively, of their problem. The author originally posed Propositions 1 and 2 as challenge problems for the students in his Math 25 class at Harvard University in Fall 2000.

A related game (the "Choquet game") can be used to prove the Baire category theorem (see [5, §8.C] and [3]). In Choquet's game, played in a given metric space X, Pierre moves first by choosing a nonempty open set $U_1 \subseteq X$. Then Paul moves by choosing a nonempty open set $V_1 \subseteq U_1$. Pierre then chooses a nonempty open set $U_2 \subseteq V_1$, and so on, yielding two decreasing sequences U_n and V_n of nonempty open sets with $U_n \supseteq V_n \supseteq U_{n+1}$ for all n, and $\bigcap U_n = \bigcap V_n$. Pierre wins if $\bigcap U_n = \varnothing$, and Paul wins if $\bigcap U_n \neq \varnothing$. One can show that if X is *complete* (i.e., if every Cauchy sequence converges in X), then Paul has a winning strategy, and if X contains a nonempty open set O that is a countable union of closed sets having empty interior, then Pierre has a winning strategy. As a consequence, one obtains the *Baire category theorem:* If X is a complete metric space, then no open subset of X can be a countable union of closed sets having empty interior.

Another related game is the Banach-Mazur game (see [7, §6] and [5, §8.H]). A subset S of the unit interval $[0,1]$ is fixed, and then Anna and Bartek alternate play. First Anna chooses a closed interval $I_1 \subseteq [0,1]$, and then Bartek chooses a closed interval $I_2 \subseteq I_1$. Next, Anna chooses a closed interval $I_3 \subseteq I_2$, and so on. Together the players' moves determine a nested sequence I_n of closed intervals. Anna wins if $\bigcap I_n$ has at least one point in common with S; otherwise Bartek wins. It can be shown (see [7, Theorem 6.1]) that Bartek has a winning strategy if and only if S is *meagre*. (A subset of X is called *nowhere dense* if the interior of its closure is empty and is called *meagre*, or of the *first category*, if it is a countable union of nowhere dense sets.) It can also be shown, using the axiom of choice, that there exist sets S for which the Banach-Mazur game is undetermined (i.e., neither player has a winning strategy).

For a more thorough discussion of these and many other *topological games*, we refer the reader to the survey article [9], which contains an extensive bibliography. Many of the games discussed in [9] are not yet completely understood.

Games like the ones we have discussed play a prominent role in the modern field of *descriptive set theory*, most notably in connection with the *axiom of determinacy* (AD). (See [4, Chapter 6] for a more detailed discussion.) Let X be a given subset of the space ω^ω of infinite sequences of natural numbers, and consider the following game between Alice and Bob. Alice begins by playing a natural number, then Bob plays another (possibly the same) natural number, then Alice again plays a natural number, and so on. The resulting sequence of moves determines an element $x \in \omega^\omega$. Alice wins if $x \in X$, and Bob wins otherwise. The axiom of determinacy states that this game is determined (i.e., one of the players has a winning strategy) for *every* choice of X.

A simple construction shows that the axiom of determinacy is inconsistent with the axiom of choice. On the other hand, with Zermelo-Fraenkel set theory plus the axiom of determinacy (ZF+AD), one can prove many nontrivial theorems about the real numbers, including that (i) every subset of $\mathbb{R}$ is Lebesgue measurable and (ii) every uncountable subset of $\mathbb{R}$ contains a perfect subset. Although ZF+AD is not considered a "realistic" alternative to ZFC (Zermelo-Fraenkel + axiom of choice), it has stimulated a lot of mathematical research, and certain variants of AD are taken rather seriously. For example, the axiom of *projective determinacy* is intimately connected with the continuum hypothesis and the existence of large cardinals (see [10] for details).

Acknowledgments. The author was supported by NSF Grant DMS-0300784.

This article was originally published in *Mathematics Magazine*, vol. 80, no. 5 (December 2007), pp. 377–380. Reprinted with permission.

Bibliography

[1] K. Ciesielski. *Set Theory for the Working Mathematician*, London Mathematical Society Student Texts 39. Cambridge, UK: Cambridge University Press, 1997.

[2] J. W. Grossman and B. Turett. "Problem #1542." *Mathematics Magazine* 71:1 (1998), 67.

[3] F. Hirsch and G. Lacombe. *Elements of Functional Analysis*, Graduate Texts in Mathematics 192. Berlin: Springer-Verlag, 1999.

[4] A. Kanamori. *The Higher Infinite*, second edition. Berlin: Springer-Verlag, 2003.

[5] A. Kechris. *Classical Descriptive Set Theory*. Berlin: Springer-Verlag, 1995.

[6] W. A. Newcomb. "Solution to Problem #1542." *Mathematics Magazine* 72:1 (1999), 68–69.

[7] J. Oxtoby. *Measure and Category*, second edition. Berlin: Springer-Verlag, 1980.

[8] W. Rudin. *Principles of Mathematical Analysis*, third edition. New York: McGraw-Hill, 1976.

[9] R. Telgársky. "Topological Games: On the 50th Anniversary of the Banach-Mazur Game." *Rocky Mountain J. Math.* 17 (1987), 227–276.

[10] H. Woodin. "The Continuum Hypothesis, Part I." *Notices of the AMS* 48:6 (2001), 567–576.

The Cyclic Butler University Game

Aviezri S. Fraenkel

In a previous book honoring Gardner, a two-player coin-pushing game on a directed graph (=digraph) without cycles was solved. The coins are placed on selected nodes of the "Butler University map." A move consists of choosing a coin and pushing it to an adjacent node along a directed edge. The player making the last move wins. We consider the same game, but where the digraph may be cyclic. Then there need not be a last move, in which case the outcome is a (dynamic) draw, that is, no player can force a win, but both always have a nonlosing next move. We provide an efficient strategy, consisting of deciding, for every position: (i) who can win, or (ii) whether both can only draw; and (iii) determining the next move that guarantees a win (case (i)) or maintains a draw (case (ii)).

In [4], Rebecca Wahl analyzed "The Butler University Game," created by Jerry Farrell. This two-player game is played on a simplified map of the Butler University campus, see Figure 1. The labels are the first letters of the various campus buildings. For example, G stands for Gallahue. Wahl explained that the roads

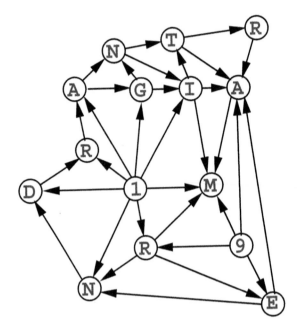

Figure 1. A simplified map of the Butler University campus.

between the buildings are directed, because a severe snowstorm led the grounds crews to mark the roads to be one-way only. Thus the map became a *directed graph*, also called, for short, *digraph.*

The game begins by placing coins on some or all of the nodes (buildings). A move consists of selecting a single coin and pushing it to an adjacent node along a directed edge, but only in the direction of the arrow. Multiple occupancy on the nodes is permitted, both in the initial placement and as the result of a move. The players move alternately. The player making the last move wins, and the opponent loses. Note that M is a *leaf*, that is, a node without any outgoing edge. So if all coins are in M, play ends.

It's not immediately clear, however, that every play of the game ends in a finite number of moves. Perhaps a player can cycle through nodes, avoiding ever entering M. But Figure 2, which is a redrawing of Figure 1, shows that the digraph is *acyclic*, i.e., without cycles. So indeed every play ends with a clear winner and loser. Incidentally, Figure 2 also shows that Butler University paid

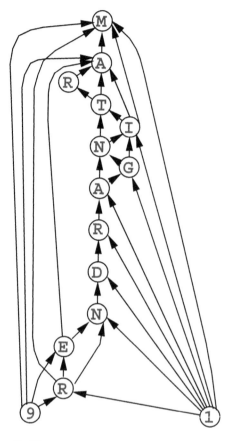

Figure 2. The snowed-in campus map is acyclic.

special tribute to Martin Gardner in his 91st year of life. It is to be hoped that, in at most another nine years, the University will complete its *Gardner Building*, so the digraph can accommodate the number 100. Using the *Sprague-Grundy function*, the Butler University game can be solved efficiently, i.e., a winning strategy can be computed "in polynomial time" [1, Section 2].

The purpose of this note is to extend these results to *cyclic* digraphs—digraphs that contain cycles. Then there may indeed be cases where play of the game doesn't end: no player can force a win, but each always has a nonlosing next move. We then say that the outcome is a *draw*. Consider, for example, the case where the snow begins to thaw just a bit, so the road between E and N_2 is back to two-way traffic (see Figure 3).

If there is a single coin at E or at N_2, the player venturing out of the cycle $(E, N_2) - (N_2, E)$ can be beaten by the opponent. Indeed, if Alice moves $E \to A_1$, then Bob moves $A_1 \to M$, winning. If Alice moves $N_2 \to D$, then Bob moves to R_2, Alice to A_2, and Bob chooses to move to N_1. If now Alice moves to I, then Bob wins immediately by going to M. Otherwise Alice moves to T. Then Bob chooses to move to R_1, Alice to A_1, Bob to M, winning.

Question 1. If coins are placed on some other nodes, in addition to a single coin at E or N_2, does the outcome always remain a draw, or may it become a win for the first player?

Every finite cyclic game can be analyzed using the *generalized Sprague-Grundy function* [3]. We demonstrate on the Butler University digraph how to compute this function.

Label M by 0. Do the same for every unlabeled node that doesn't have an (immediate) *follower* labeled 0, and every unlabeled follower of which has a follower labeled 0. Therefore, R_1 is labeled 0. Then N_1 is labeled 0, since the follower I has the follower M labeled 0, and the follower T has the follower R_1 labeled 0. Then R_2 is labeled 0. The next candidate is N_2 since its unlabeled follower D has the follower R_2 already labeled 0. However, it has also the unlabeled follower E, which doesn't have any follower labeled 0. Thus, N_2 cannot be labeled 0. Also no other unlabeled node can be labeled 0.

At this stage, every unlabeled node that doesn't have a follower 0 is labeled with a special symbol ∞, larger than every integer. Therefore we dub it *infinity*. This implies that E and N_2 are labeled ∞. But the nodes R_3, 1, and 9 remain unlabeled at this stage, since each has a follower labeled 0.

We now label by 1 every unlabeled node that doesn't have a follower 1, and every follower that is either unlabeled or labeled ∞ has a follower labeled 1. Thus A_1 is labeled 1, then G and D are labeled 1, so also R_3 is labeled 1. Every unlabeled node now has a follower 1, so no additional ∞ labels are dispensed.

Repeat this procedure with the label 2. Thus T is labeled 2, then A_2. Now node 9 has the follower ∞ (at E), which doesn't have a follower labeled 2, hence node 9 is labeled ∞. Then I is labeled 3. Node 1 is a candidate for 4, but it has a follower ∞ (at N_2), which doesn't have a follower labeled 4, so node 1 is labeled ∞. Finally,

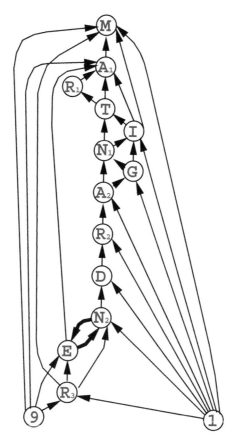

Figure 3. The snow begins to thaw.

we adjoin to every ∞-label the set of labels of all its finite followers. The labeled digraph is depicted in Figure 4.

We now show how to use the labels for playing the game optimally. Suppose that one coin is placed on each of the nodes E, R_2, and I, indicated by stars in Figure 4. We use *Nim-addition* to add the labels of the occupied nodes as follows:

$$\infty(1) \oplus 0 \oplus 3 = \infty(1) \oplus 3 := \infty(1 \oplus 3) = \infty(2).$$

We used $\oplus$ to indicate Nim-addition, which is binary addition without carries. Thus $1 \oplus 3$ (in decimal) $= 01 \oplus 11$ (in binary) $= 10$ (in binary) $= 2$ (in decimal). In fact, there are only 10 types of people

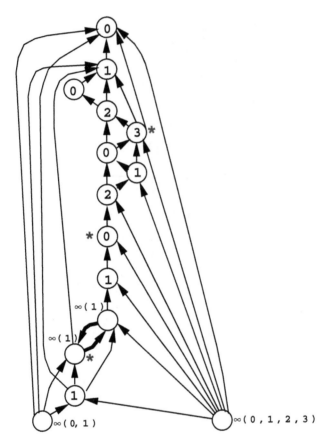

Figure 4. A fully labeled Butler University cyclic digraph.

in the world: those who understand binary, and those who don't. Since the Nim-sum is $\infty(2)$, which doesn't contain 0, the position is a draw. The way to maintain the draw is by moving in the cycle. If a player moves from a coin on a node labeled ∞ to a node labeled 1, then the opponent can win by moving the coin on the node labeled 3 to a node labeled 1. Then the Nim-sum of the resulting position is $0 \oplus 1 \oplus 1 = 1 \oplus 1 = 0$. Every position whose Nim-sum is 0 is a second-player win.

Does this indicate that the answer to Question 1, that placing coins on other nodes in addition to a coin on a node labeled ∞, implies that the outcome necessarily remains a draw?

Well, suppose that the initial placement of coins is as above, except that the coin on the node labeled 0 is shifted one node down, namely, it is placed initially on the node labeled 2. Then the Nim-sum of the initial position is

$$\infty(1) \oplus 2 \oplus 3 = \infty(1) \oplus 1 = \infty(1 \oplus 1) = \infty(0).$$

Since there is a 0 attached to the ∞, the first player can win by moving from the node labeled ∞ to a node labeled 1. Indeed, the resulting Nim-sum is then $1 \oplus 2 \oplus 3 = 0$, which is a second-player win.

In conclusion, the answer to Question 1 is that the positions of the additional coins determine whether the outcome remains a draw or becomes a win for the first player.

Now suppose that the snow has melted some more, and a loop around building A_2 (Figure 3) became passable, in addition to the two-way road opened up before. Note that at A_2 there is now the option of *passing*: a player moving around the loop stays in the same position!

Question 2. Is it true that then every play with a coin on A_2, where there is a loop, is a draw?

Figure 5 depicts the situation with a coin on A_2 and two additional coins, indicated by stars, as well as the new labeling induced by the loop. The Nim-sum of the occupied nodes is now $\infty(0,1) \oplus 2 \oplus 3 = \infty(0,1) \oplus 1 = \infty(1,0)$. Thus, the first player can win by moving to 1. Indeed, the Nim-sum of this position is $1 \oplus 2 \oplus 3 = 0$. Note, in fact, that if there is only a single coin at A_2, then the first player can win by moving to the follower labeled 0. However, if the initial position were to have only the coin on A_2 and precisely *one* of the other two coins depicted by stars in Figure 5, then the result would be a draw. Thus, also the answer to Question 2 is that the outcome depends on the placement of the coins.

The following is a summary of the main steps of the generalized Sprague-Grundy function labeling algorithm, for any finite digraph.

1. Put $i = 0$.

2. As long as there exists an unlabeled node u such that no follower of u is labeled i and every follower of u that is either unlabeled or labeled ∞ has a follower labeled i, label u by i.

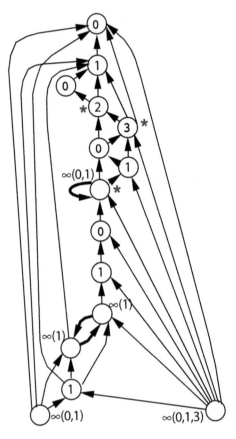

Figure 5. A fully labeled Butler University cyclic digraph, including a loop.

3. Label every unlabeled node that doesn't have a follower labeled i by ∞.

4. If there is still an unlabeled node, put $i \leftarrow i+1$ (i.e., increase
   i by 1) and return to 2. Otherwise stop.

A simple mathematical definition of the generalized Sprague-
Grundy function can be found in [1, Section 3]. The complete labeling algorithm can be found in [2, Section 3].

We end with a homework problem about multiple occupancy.
Suppose that finitely many coins are placed on a single node u of

a finite digraph, all other nodes being unoccupied. Denote by $\ell(u)$ the label at u. Prove the following:

(i) If $\ell(u) = 0$, then no matter how many coins are placed on u, the outcome is a second player win.

(ii) If $0 < \ell(u) < \infty$, then the outcome is a second player win if the number of coins on u is even. Otherwise it's a first player win.

(iii) If $\ell(u) = \infty$, then the outcome is a draw for any number of coins ≥ 2 on u.

Bibliography

[1] A. S. Fraenkel. "Scenic Trails Ascending from Sea-Level Nim to Alpine Chess." In *Games of No Chance*, edited by R. J. Nowakowski, Mathematical Sciences Research Institute Publications 29, pp. 13–42. Cambridge, UK: Cambridge University Press, 1996.

[2] A. S. Fraenkel. "Recent Results and Questions in Combinatorial Game Complexities." *Theoretical Computer Science* 249 (2000), 265–288.

[3] C. A. B. Smith. "Graphs and Composite Games." *Journal of Combinatorial Theory* 1 (1966), 51–81.

[4] R. G. Wahl. "The Butler University Game." In *Tribute to a Mathemagician*, edited by B. Cipra, E. D. Demaine, M. L. Demaine and T. Rodgers, pp. 37–40. Wellesley, MA: A K Peters, 2005.

Misere Play of G-A-R-D-N-E-R, the G4G7 Heptagon Game

Thane Plambeck

Martin Gardner treated two-player impartial combinatorial games several times in his *Scientific American* columns. Such a game is exemplified by the game of nim, which can be played with heaps of beans. On a player's move in nim, he is allowed to remove as many beans as he likes from a single heap (including the whole heap, if desired). In normal play, the last player to remove the final bean from the final remaining heap is declared the winner of the game. In misere play, that player loses the game.

Misere play of impartial games is typically much more difficult to analyze. In this article we illustrate how to analyze a simple misere coin-sliding game called G-A-R-D-N-E-R using a theory that generalizes the Sprague-Grundy theory for normal play.

The game board is shown in Figure 1.

The G-A-R-D-N-E-R game is based on one of the logos of the Gathering for Gardner 7 conference, which is shown in Figure 2.

There are two players. To set up the board for play, stacks of coins are placed on various vertices of the game board, excluding the topmost "G" node. It doesn't matter how many coins are placed

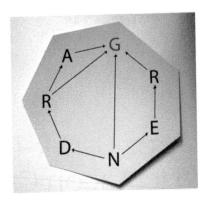

Figure 1. The game board for G-A-R-D-N-E-R.

on the board, but there should be at least six to ten coins, and they should be scattered amongst the various vertices. For example, one might start with one, two, or three coins at each non-G vertex.

On a player's move, he slides one coin along a single directed edge of the game board. The direction of the arrows ensures that all coins will eventually reach the "sink" node "G" at the top of the board.

Normal Play

As a warm-up, let's work out a complete winning strategy for *normal play* of G-A-R-D-N-E-R. In normal play, the last player to slide a coin to the "G" node is declared the *winner* of the game. We can describe a simple strategy for normal-play G-A-R-D-N-E-R using the *Sprague-Grundy theory*. Figure 3 illustrates the *nim-heap equivalents* for each vertex of the board.

Figure 2. A Gather for Gardner 7 logo.

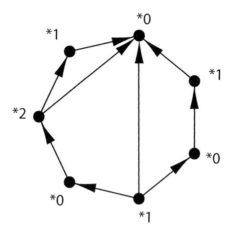

Figure 3. The nim-heap equivalents of the G-A-R-D-N-E-R board vertices.

The values $*0$, $*1$, $*2$, and $*3$ are added using the rule of *nim addition* (binary addition without carrying):

+	$*0$	$*1$	$*2$	$*3$
$*0$	$*0$	$*1$	$*2$	$*3$
$*1$	$*1$	$*0$	$*3$	$*2$
$*2$	$*2$	$*3$	$*0$	$*1$
$*3$	$*3$	$*2$	$*1$	$*0$

A given normal-play position is a winning position for the next player to move (a so-called *N-position*) if and only if when we sum up the nim values of each coin, the result is $*1$, $*2$, or $*3$. Positions that add up to $*0$ are winning for the second player to move (*P-positions*).

For example, suppose two coins are at vertex "N." The nim sum

$$*1 + *1 = *0$$

says that the position is a P-position in normal play.

Misere Play

In *misere play*, the last player to slide a coin to the "G" node is declared the *loser* of the game. Is it possible to give a similar assignment of values to the single vertices of the G-A-R-D-N-E-R board

so that we can determine who wins in misere play by "adding" up
values? The answer is yes, but the desired values are not nim-
heaps, and the addition is not nim-addition. Instead, we assign to
the vertices of the board values taken from a particular commutative 14-element monoid Q, the *misere indistinguishability quotient*
of G-A-R-D-N-E-R.

A *monoid* is a semigroup with an identity. To write down the addition of misere G-A-R-D-N-E-R, we can use a *commutative monoid
presentation*:

$$Q = \langle a, b, c \mid a^2 = 1, b^3 = b, b^2c = c, c^3 = ac^2 \rangle.$$

A general monomial of the form

$$a^i b^j c^k$$

can always be reduced to one of the 14 elements of Q:

$$Q = \{1, a, b, c, ab, ac, b^2, ab^2, bc, abc, abc^2, c^2, ac^2, bc^2\}.$$

The 14 elements of Q partition into two subsets

$$Q = P \bigcup N,$$

where

$$P = \{a, b^2, c^2, bc\}$$

are the P-positions and

$$N = \{1, b, c, ab, ac, ab^2, abc, abc^2, ac^2, bc^2\}$$

are the N-positions.

To win at misere G-A-R-D-N-E-R from a given N-position, we
need to a move to a P-position. This is simplified by using the
graphical information provided in Figure 4. (It also shows the assignment quotient elements to the game board vertices, in the upper right hand corner).

Finding a Move in an Example Misere
G-A-R-D-N-E-R Position

Consider, for example, the misere G-A-R-D-N-E-R position P consisting of exactly one coin at each of the non-"G" vertices of the

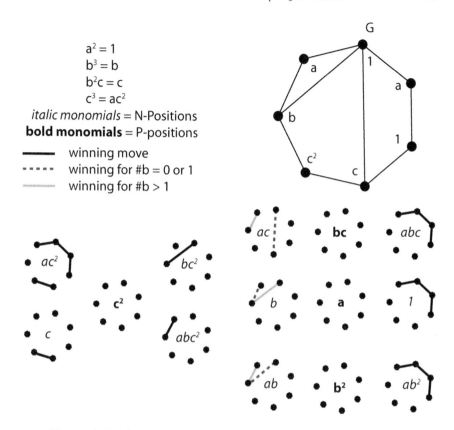

Figure 4. Finding a winning position for misere G-A-R-D-N-E-R.

game board. Moving counterclockwise from the "G" around the diagram shown in the upper right-hand corner of Figure 4, the corresponding single-coin monomials are

$$a, b, c^2, c, 1, a.$$

To determine the monomial for the multi-coin position P, we multiply these corresponding single-coin monomials together, apply commutativity, and reduce the resulting monomial via the four G-A-R-D-N-E-R relations (shown in gray). The product of the six monomials above is

$$a \cdot b \cdot c^2 \cdot c \cdot 1 \cdot a = a^2 bc^2,$$

which can then be reduced via the relations $a^2 = 1$ and $c^3 = ac^2$ to obtain the monomial

$$abc^2,$$

which cannot be further reduced using the four G-A-R-D-N-E-R relations. Again consulting the diagram, we see that abc^2 is one of the nonbold monomials in the diagram (i.e., an N-position). The original position P, therefore, has a forced winning move for the first player. The winning move

$$R \rightarrow A$$

is shown as a black edge in the (small) abc^2 monomial diagram in Figure 4. In terms of single-coin monomials, the winning move is a transition

$$a \rightarrow b.$$

Next, let's check that the resulting position, after the recommended move has been made, is indeed a P-position. After the winning move has been made, the resulting position has two coins at "A," none at the "R" labeled b, and one each at each other non-G vertex. The corresponding result monomial is

$$a^2 \cdot c^2 \cdot c \cdot 1 \cdot a = a^3 c^3,$$

which reduces to the bold monomial (i.e., P-position)

$$ac^3 = a \cdot a \cdot c^2 = c^2.$$

For more information on the indistinguishability quotient construction, see [1] and [2]. To make the game board and print the stickers that go on the back, see the following links:

- Front board (print it on card stock and cut on lines): http://www.plambeck.org/archives/heptboard-boardonly.pdf

- Sticker sheet 1 (suitable for 4-up printing [5168 Avery labels]): http://www.plambeck.org/archives/stickers-heptagon-FINAL-1.pdf

- Sticker sheet 2 (suitable for 14-up printing [5262 Avery labels]): http://www.plambeck.org/archives/stickers-heptagon-FINAL-2.pdf

Bibliography

[1] Thane Plambeck. "Taming the Wild in Impartial Combinatorial Games." arXiv:math/0501315v2, 2005. Available at http://arxiv.org/abs/math.CO/0501315.

[2] Thane Plambeck. "Advances in Losing." arXiv:math/0603027v1, 2006. Available at http://arxiv.org/abs/math.CO/0603027.

Part IV

Play with Numbers

The Association Method for Solving Certain Coin-Weighing Problems

Dick Hess

Coin-weighing problems have been popular in recreational mathematics for at least the past 60 years. Certain such problems, in which a counterfeit coin or coins must be identified through weighing, can be solved by a particularly useful and effective method I call the *Association Method*. This method is described and examples are given for problems involving three kinds of weighing devices:

1. A simple two-pan balance.

2. A one-pan pointer scale.

3. A two-pan pointer scale, which reads the signed difference in weights between whatever (if anything) is placed on the two pans.

In all problems, true and fake coins look and feel identical; the only way to distinguish them from each other is by weighing, using the device provided.

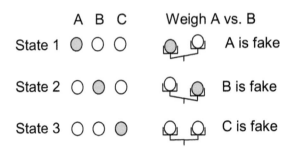

Figure 1. Possible outcomes for the three coin problem.

A Simple Coin Problem

Imagine a collection of three identical looking coins (A,B,C) of which
two are true and known to weigh the same and one is counterfeit
and slightly heavier than the other two. With a simple two-pan balance determine which coin is counterfeit in one weighing. Figure 1
demonstrates a way to solve the problem by weighing coin A vs.
coin B and examining the outcomes for the three possible states
as shown.

The Association Method

The Association Method considers all possible weighing instructions for a coin. In the simple problem above, there are three possible instructions for any one coin:

> **L**: Put the coin on the left pan of the balance.
> **R**: Put the coin on the right pan of the balance.
> **0**: Do not weigh the coin.

The Association Method associates these instructions with the
coins in a one-to-one fashion. Table 1 shows such an association
and how it identifies the fake coin for all possible outcomes.

Associations	Outcomes
A: **L**	If left pan is heavy, then A is fake.
B: **R**	If right pan is heavy, then B is fake.
C: **0**	If there is balance, then C is fake.

Table 1. The Association Method for a simple three coin problem.

Associations	Outcomes
A: (**L,L**); D: (**R,R**); G: (**0,R**)	If L is heavy in W1 and L is heavy in W2, then A is fake.
B: (**L,R**); E: (**0,L**); H: (**R,0**)	If L is heavy in W1 and R is heavy in W2, then B is fake.
C: (**R,L**); F: (**L,0**); I: (**0,0**)	If R is heavy in W1 and L is heavy in W2, then C is fake, etc.

Table 2. The Association Method for a nine coin problem, where L = left pan, R = right pan, W1 = first weighing, and W2 = second weighing.

Expansion to More Weighings

Suppose the problem is made more complicated by involving nine coins (A to I) of which eight coins are true and one coin is a slightly heavy counterfeit. For this problem, the challenge is to find the fake in two weighings. A solution to this problem using the Association Method is demonstrated in Table 2. In this case, there are nine different sets of instructions to be associated with the nine coins. Each instruction set gives directions on what to do with the coin in each of the two weighings. For example, the instruction set (**L,0**) says to put the coin associated with it on the left pan for the first weighing and not to involve it in the second weighing.

There are several advantages of the Association Method. First, it is automatic in that the fake coin can be read off directly from the outcomes of the weighings. Second, it produces a completely predetermined or oblivious decision tree in which subsequent weighings don't depend on the outcomes of prior weighings. Third, it is optimal for predetermined decision trees in that no additional coin can be accommodated for the same number of weighings. This follows because all possible sets of weighing instructions are employed; any additional coin would have to duplicate a set and thus be indistinguishable from the coin already associated with that set of weighing instructions. The problem above can be generalized to 27, 81, 243, . . . coins in 3, 4, 5, . . . weighings.

A Modified Problem

Suppose two weighings are still allowed to find the counterfeit but we don't know if the coin is heavier or lighter than a true coin. How many coins can be allowed in this case? As before, there are

Associations	Outcomes
A: **(L,R)**	**(+,−)** means A is heavy; **(−,+)** means A is light.
B: **(R,0)**	**(−,0)** means B is heavy; **(+,0)** means B is light.
C: **(0,L)**	**(0,+)** means C is heavy; **(0,−)** means C is light.
D: **(0,0)**	**(0,0)** means D is fake (not known whether heavy or light).

Table 3. The Association Method for a four coin problem.

nine different instructions sets, but two restrictions arise. First, all but the **(0,0)** instruction set have a partner that must be discarded to avoid ambiguity. The partner of an instruction set is that instruction set with **L** and **R** interchanged. For example, **(L,0)** has a partner **(R,0)**. If both are included for association, we can't distinguish between a heavy coin associated with **(L,0)** and a light coin associated with **(R,0)**. Thus, we are down to five allowable instruction sets to associate. The second restriction is that each weighing must have the same number of coins on each side of the balance. If all five allowable instruction sets are chosen, each weighing will not have the same number of coins in each pan of the balance. Thus one more instruction set must be discarded. The best case is shown in Table 3, which indicates how to find a counterfeit among four coins (A, B, C, and D). Outcomes of the two weighings are indicated, for example, by **(+, −)** if the left pan is heavy on the first weighing and light on the second weighing, and so on. This approach generalizes to more weighings, demonstrating that a counterfeit can be determined among 13, 40, 121, ... coins if 3, 4, 5, ... weighings are allowed.

The Classic 12 Coin Problem

In the classic 12 coin problem (first published and analyzed about 60 years ago) there are 12 coins (A to L) of which one is counterfeit. Three weighings are permitted to identify the counterfeit and determine if it is heavier or lighter than a true coin. The Association Method elegantly solves this problem automatically, as shown in Table 4. It is the same as the 13 coin problem alluded to in the prior section but the instruction set **(0,0)** must be excluded so we can determine if the fake is heavy or light in all cases.

Associations	Outcomes	Implied Weighings
A: (**L,L,R**)	(**+,+,−**) means A is heavy; (**−,−,+**) means A is light	
B: (**L,R,R**)	(**+,−,−**) means B is heavy; (**−,+,+**) means B is light	W1: ABGJ vs. CEHI
C: (**R,L,R**)	(**−,+−**) means C is heavy; (**+,−,+**) means C is light	W2: ACIK vs. BDFH
D: (**0,R,L**)	(**0,−,+**) means D is heavy; (**0,+,−**) means D is light	W3: DEGL vs. ABCF
E: (**R,0,L**)	(**−,0,+**) means E is heavy; (**+,0,−**) means E is light	
F: (**0,R,R**)	(**0,−,−**) means F is heavy; (**0,+,+**) means F is light	
G: (**L,0,L**)	(**+,0,+**) means G is heavy; (**−,0,−**) means G is light	
H: (**R,R,0**)	(**−,−,0**) means H is heavy; (**+,+,0**) means H is light	
I: (**R,L,0**)	(**−,+,0**) means I is heavy; (**+,−,0**) means I is light	
J: (**L,0,0**)	(**+,0,0**) means J is heavy; (**−,0,0**) means J is light	
K: (**0,L,0**)	(**0,+,0**) means K is heavy; (**0,−,0**) means K is light	
L: (**0,0,L**)	(**0,0,+**) means L is heavy; (**0,0,−**) means L is light	

Table 4. The Association Method for the classic 12 coin problem.

Two Problems Using a One-Pan Pointer Scale

Suppose we are presented with 25 stacks of four coins each from 25 different mints and are told that one mint consistently produces fake coins, each weighing 11 grams instead of the 10 gram weight of a true coin. We are allowed to use an accurate one-pan pointer scale to find the fakes in two weighings. The Association Method is ideal for solving this problem as follows. There are 25 possible doublets, (a,b), where a = 0, 1, 2, 3, or 4 and b = 0, 1, 2, 3, or 4. Each doublet can be interpreted as a weighing instruction set and associated uniquely to one of the stacks. Table 5 demonstrates one such way. The implied weighings each involve 50 coins and would both read 500 grams if all coins were true. The amounts over 500 grams for the two weighings then directly identify the

Associations				
A: (0,0)	F: (1,0)	K: (2,0)	P: (3,0)	U: (4,0)
B: (0,1)	G: (1,1)	L: (2,1)	Q: (3,1)	V: (4,1)
C: (0,2)	H: (1,2)	M: (2,2)	R: (3,2)	W: (4,2)
D: (0,3)	I: (1,3)	N: (2,3)	S: (3,3)	X: (4,3)
E: (0,4)	J: (1,4)	O: (2,4)	T: (3,4)	Y: (4,4)

Table 5. Associations for the problem of 25 stacks of four coins each.

mint producing the fake coins. For example, if W1 = 503 grams and W2 = 502 grams, then the doublet (3,2) is determined and mint R is producing counterfeit coins. The method is expandable to $N = (n+1)^w$ stacks of n coins each if w weighings are allowed and the weights of true and false coins are known.

Now suppose the problem is modified so that true coins are known to weigh 10 grams each but coins in the false stack are all the same but of unknown weight. If there are N stacks of four coins each, what is the largest possible value of N that allows us to find the bad mint in two weighings? The Association Method can still be applied but must be modified to tackle the problem. Of the 25 weighing instruction sets presented in the prior case, some must be removed to eliminate ambiguity. Any two that are multiples of each other will produce such an ambiguity because we don't know the weight difference between a true and false coin. Thus, one of these two weighing instruction sets must be removed. Table 6 shows one way to remove all such multiples, leaving 14 weighing instruction sets. Both implied weighings involve 26 coins and the proportion (W1-260):(W2-260) determines the mint producing fakes. For example, if W1 = 251 grams and W2 = 254 grams then (W1-260):(W2-260) = 3:2, the weighing instruction set (3,2) is determined, and mint K is producing counterfeit coins. For

Associations				
A: (0,0)	C: (1,0)	(2,0)	(3,0)	(4,0)
B: (0,1)	D: (1,1)	H: (2,1)	J: (3,1)	M: (4,1)
(0,2)	E: (1,2)	(2,2)	K: (3,2)	(4,2)
(0,3)	F: (1,3)	I: (2,3)	(3,3)	N: (4,3)
(0,4)	G: (1,4)	(2,4)	L: (3,4)	(4,4)

Table 6. How the fakes can be determined from 14 stacks of four coins each.

n	$w = 2$	$w = 3$	$w = 4$	$w = 5$	$w = 6$
1	4	8	16	32	64
2	6	20	66	212	666
3	10	50	226	962	3970
4	14	92	530	2852	14834
5	22	176	1186	7472	45802
6	26	254	2066	15542	112826
7	38	416	3746	31472	257258
8	46	572	5842	55652	515026
9	58	806	9106	95822	980218
10	66	1034	13026	152042	1720146
11	86	1424	19106	239792	2934506
12	94	1724	25522	351332	4693474

Table 7. Number of distinguishable stacks for a one-pan pointer scale.

this case it would also be deduced that counterfeit coins weigh 7 grams each. The approach generalizes for n coins per stack and w weighings. Table 7 shows the values of N for small n and w.

A Final Puzzle Using a Two-Pan Pointer Scale

Now consider N stacks of two coins each from different mints where we know only that a false coin is heavier than a true coin but don't know the weight of either type of coin. Again there is one stack of false coins, each weighing the same unknown amount. How large can N be to allow us to find the false coins in three weighings on a two-pan pointer scale? This weighing device reads the signed difference in weights between whatever (if anything) is placed on the two pans. The Association Method solves this problem beautifully in two steps. First, list all triplets of nonnegative integers 0,1,2 that are not multiples of each other:

> 000;
> 100, 010, 001;
> 011, 101, 110, 012, 102, 120, 021, 201, 210;
> 112, 121, 211, 122, 212, 221, 111.

They fall as shown into four groups, depending upon how many positive entries are in each triplet. The next step is to form all possible weighing instruction sets from these triplets. The triplet 100

n	$w = 2$	$w = 3$	$w = 4$	$w = 5$	$w = 6$
1	9	27	81	243	729
2	17	99	545	2883	14897
3	33	291	2241	16323	116193
4	49	579	5857	55683	515089
5	81	1155	13857	157443	1754481
6	97	1731	25537	351363	4693537
7	145	2883	47521	739203	11256625
8	177	4035	76257	1357443	23589777
9	225	5763	121281	2400003	46396065
10	257	7491	176897	3903363	83862017
11	337	10371	262177	6255363	146131057
12	369	12675	355425	9335043	238781649

Table 8. Number of distinguishable stacks for a two-pan pointer scale. All coins are of unknown weight. Fake coins are heavier than true coins.

and those from its group can each form two weighing instruction sets: (L,0,0) and (R,0,0) from 100. The triplet 102 and those from its group can each form four different weighing instruction sets: (L,0,2L), (L,0,2R), (R,0,2L), and (R,0,2R) from 102. The triplet 112 and those from its group can each form eight different weighing instruction sets: (L,L,2L), (L,L,2R), ..., (R,R,2R) from 112. With the weighing instruction set (0,0,0) there is a total of $1 + 2 \times 3 + 4 \times 9 + 8 \times 7 = 99$ possible different weighing instruction sets. Thus, 99 stacks of two coins can be distinguished in three weighings on a two-pan pointer scale. This approach generalizes for n coins per stack and w weighings. Table 8 shows the values of N for small n and w.

Conclusion

The Association Method is elegant and effective in solving certain coin-weighing problems as demonstrated above. It can be applied to problems involving at least three different types of weighing devices and has some significant advantages over many other methods. First, it is automatic in that the fake coin or stack of coins can be directly read off from the outcomes of the weighings. Second, it produces a completely predetermined or oblivious decision tree in which subsequent weighings don't depend on the outcomes of prior weighings. Third, it is optimal for predetermined decision trees in that no additional coin or stack of coins can be accommodated for the same number of weighings.

The Art of Ready Reckoning

Mogens Esrom Larsen

Recall Dustin Hoffman, who—as *Rainman*—throws numbers out of his sleeves. He looks at a box of matches dropped on the floor and immediately claims "246" (or whatever).

I feel rather like Alice in *Through the Looking Glass and What Alice Found There*, Chapter IX (Lewis Carroll, 1871):

> "Can you do Addition?" the White Queen asked.
>
> "What's one and one and one and one and one and one and one and one and one and one?"
>
> "I don't know," said Alice, "I lost count."
>
> "She can't do Addition," the Red Queen interrupted.

I do not believe that autistic people have special gifts for dealing with numbers. But if they are interested in numbers, their capacity for monomania allows them to play with them day and night, so that numbers become their personal friends. In this respect they are like the famous Indian mathematician, S. Ramanujan (1887–1920). According to a well-known anecdote, G. H. Hardy (1877–1947) once remarked to Ramanujan: "I think the number of my taxicab was 1729; it seemed to me a rather dull number." Ramanujan answered, "No, Hardy! It is a very interesting number. It is the smallest number expressible as the sum of two cubes in two different ways."

$$1729 = 10^3 + 9^3 = 12^3 + 1^3$$

Playing with numbers, you note those kinds of coincidences.

Forty years ago, the main question was whether you could reckon without using your head. Now we are so used to calculators that it becomes a challenge to manage without one.

The Multiplication Table

To reckon in your head, you have to know by heart the tables of products (and sums) at least up to 9.

It is no big deal; altogether we must know 45 sums and 36 products, or 31 prime resolutions.

In this requirement we are lucky. In ancient Assyria or Babel a couple of thousand years B.C., they used a positional system with base 60 (we still have a reminder of that system in our time units of minutes and seconds), so they needed to know the products up to 59. Remembering the rule of commutation this results in

$$\frac{58 \times 57}{2} + 58 = 30 \times 57 + 1 = 1711$$

products to remember.

The Multiplication Table over Ten

Multiplying a one-digit number by a two-digit number is hard to simplify. For example, $7 \times 17 = 70 + 49 = 119$. Sometimes prime resolutions may give shortcuts, e.g., $9 \times 18 = 9 \times 9 \times 2 = 81 \times 2 = 162$. But for two numbers in the tens, a rewriting may improve

$$(10 + x)(10 + y) = (10 + x + y)10 + xy,$$

e.g.,

$$13 \times 19 = (13 + 9)10 + 3 \times 9 = 220 + 27 = 247$$
$$17 \times 19 = (17 + 9)10 + 7 \times 9 = 260 + 63 = 323.$$

So, learning by heart is unnecessary.

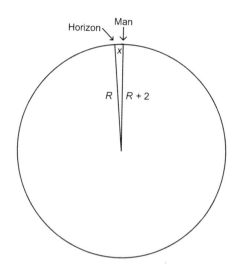

Figure 1. Measuring the distance to the horizon.

Without the Pocket Calculator

Think of yourself on the beach looking out over the ocean, in a swimsuit and hence without a pocket for a calculator, wondering, how far away is the horizon? Can you calculate that in your head?

Indeed! (See Figure 1.)

The first problem is that you do not remember the radius of the earth!

But we recall from our history lessons that the meter was defined by setting the distance from the equator to the north pole equal to 10,000 km. So the circumference of the earth is by definition 40,000 km! With $\pi \approx \frac{22}{7}$, we may find the radius as

$$R = \frac{40000}{2\pi} \approx \frac{70000}{11} = 6363.$$

So, assuming that your eyes are approximately 2 m over the surface:

$$R^2 + x^2 = \left(R + \frac{2}{1000}\right)^2 = R^2 + \frac{4R}{1000} + \frac{4}{1000000}.$$

Reckoning that the square 4/1000000 is next to nothing, and observing that the the squares R^2 cancel, we get

$$x^2 = \frac{4R}{1000} = \frac{4 \times 6363}{1000} = 25.4.$$

Hence we conclude that the distance is slightly more than 5 km.

How much more? Yes, we may get closer: if we add a little (ϵ) to 5, we get the square

$$(5+\epsilon)^2 = 25 + 10 \times \epsilon + \epsilon^2,$$

so that if the last square is quite small, we may choose $\epsilon = 0.04$ and therefore the distance to the horizon in km is 5.04.

How much did we calculate? We divided 70 by 11 and multiplied 636 by 4.

Some Useful Formulas

A couple of algebraic formulas are extremely useful. They are the square of a sum of two terms and the product of a sum and a difference of the same two numbers:

$$(x+y)^2 = x^2 + 2xy + y^2,$$
$$(x+y)(x-y) = x^2 - y^2,$$

e.g.,

$$7 \times 13 = (10-3)(10+3) = 100 - 9 = 91$$

and—slightly changed—

$$x^2 = y^2 + (x+y)(x-y),$$

e.g.,

$$59^2 = 60^2 + (59+60)(59-60) = 3600 - 119 = 3481.$$

For $y = 1$ denote

$$(x+1)^2 = x^2 + 2x + 1 = x^2 + x + (x+1),$$

e.g.,

$$21^2 = 20^2 + 20 + 21 = 400 + 41 = 441,$$
$$22^2 = 21^2 + 21 + 22 = 441 + 43 = 484.$$

Squares

Numbers ending with the digit 5 are extremely easy to square:

$$(x \times 10 + 5)^2 = 5^2 + x(x+1) \times 100,$$

e.g.,

$$25^2 = 25 + 2 \times 3 \times 100 = 625,$$
$$85^2 = 25 + 8 \times 9 \times 100 = 7225,$$
$$125^2 = 25 + 12 \times 13 \times 100 = 15625.$$

Squares may also be easily calculated with a reference to 50 or 100. The following formulas may be used, if we know the squares of some smaller numbers:

$$(50 + x)^2 = 2500 + x \times 100 + x^2 = (25 + x) \times 100 + x^2,$$
$$(100 + x)^2 = 10000 + 2x \times 100 + x^2 = (100 + 2x) \times 100 + x^2.$$

If you remember or find easily

$$18^2 = 260 + 64 = 324$$

or perhaps

$$23^2 = 3^2 + (23 - 3)(23 + 3) = 9 + 20 \times 26 = 9 + 520 = 529$$
$$23^2 = (25 - 2)^2 = 625 - 100 + 4 = 529,$$

then the formulas above yield

$$54^2 = (25 + 4) \times 100 + 4^2 = 2916,$$
$$26^2 = (25 - 24) \times 100 + 24^2 = 100 + 576 = 676,$$
$$37^2 = (25 - 13) \times 100 + 13^2 = 1200 + 169 = 1369,$$
$$92^2 = (100 - 16) \times 100 + 8^2 = 8464,$$
$$86^2 = (100 - 28) \times 100 + 14^2 = 7200 + 196 = 7396,$$
$$117^2 = (100 + 34) \times 100 + 17^2 = 13400 + 289 = 13689.$$

More General Formulas

Many of the formulas above are special cases of the following surprisingly useful "triviality":

$$(a+b)(a+c) = (a+b+c)a + bc.$$

It's obvious that for $b = c$ or $b = -c$ we get some of the previous examples. But in general it is useful with a convenient choice of a, e.g., a multiple of 10 or 100.

$$23 \times 29 = 32 \times 20 + 3 \times 9 = 640 + 27 = 667$$
$$23 \times 29 = 22 \times 30 + (-7)(-1) = 660 + 7 = 667$$
$$37 \times 34 = 41 \times 30 + 7 \times 4 = 1230 + 28 = 1258$$
$$37 \times 34 = 31 \times 40 + (-3)(-6) = 1240 + 18 = 1258$$
$$106 \times 97 = (100 + 6 - 3) \times 100 + 6 \times (-3) = 10300 - 18 = 10282$$
$$705 \times 694 = (700 + 5)(700 - 6) = 699 \times 700 - 30 = 490000 - 700 - 30$$
$$= 489270$$
$$12^3 = 36 \times 48 = (40 - 4)(40 + 8) = 44 \times 40 - 4 \times 8 = 1760 - 32$$
$$= 1728$$

The Cross Product

The general case with no obvious shortcuts may be handled with the so-called cross product. The numbers xy and zw are imagined in two lines:

$$
\begin{array}{ccc}
x & & y \\
| & \times & | \\
z & & w
\end{array}
$$

Then you create the two products $xz \times 100$ and yw, to be imagined as a four-digit number. Then you create the sum $xw + yz$ to be multiplied by 10 and added, e.g.,

$$63 \times 46 = 2418 + (6 \times 6 + 3 \times 4)10 = 2418 + (36 + 12)10$$
$$= 2418 + 480 = 2898,$$
$$36 \times 48 = 1248 + (3 \times 8 + 6 \times 4)10 = 1248 + 480 = 1728.$$

If $x = z$ or $y = w$, you may make the last part easier:

$$34 \times 37 = (900 + 28) + 3(4 + 7)10 = 928 + 330 = 1258,$$
$$63 \times 43 = (2400 + 9) + 3(6 + 4)10 = 2409 + 300 = 2709.$$

To Juggle with Many Numbers

Once upon a time, the later very famous mathematician, Carl Friedrich Gauss (1777–1855), annoyed the teacher in his first-grade class by being so brilliant. To get a rest the teacher asked him to add the numbers from 1 to 100. Gauss immediately answered $5,050$.

He imagined them in two lines forward and backward:

$$
\begin{array}{cccccc}
1 & 2 & 3 & \ldots & 99 & 100 \\
100 & 99 & 98 & \ldots & 2 & 1
\end{array}
$$

We note that there are 100 sums of two numbers, each giving the result 101. So, twice the sum in question is $10,100$.

The general formula is

$$1 + 2 + \cdots n = \frac{n(n+1)}{2}.$$

If we similarly added the odd numbers we would get

$$1 + 3 = 4, \quad 4 + 5 = 9, \quad 9 + 7 = 16, \quad 16 + 9 = 25,$$

all squares. Is this a rule?

Following Gauss, we may write the sum of the first n odd numbers forward and backward:

$$
\begin{array}{ccccccc}
1 & 3 & 5 & \ldots & 2n-3 & 2n-1 \\
2n-1 & 2n-3 & 2n-5 & \ldots & 3 & 1
\end{array}
$$

and note that there are n sums of two numbers each adding to $2n$. Twice the sum is hence $2n^2$.

Formulas for the sums of powers are known for all exponents. The sum of the squares is

$$1 + 2^2 + 3^2 + \cdots + n^2 = \frac{n(n+1)(2n+1)}{6}.$$

For example, we have

$$1 + 4 + 9 + 16 + 25 + 36 = \frac{6 \times 7 \times 13}{6} = 91.$$

For the cubes we have a beauty:

$$1 + 8 = 9, \quad 9 + 27 = 36, \quad 36 + 64 = 100, \quad 100 + 125 = 225, \quad \ldots.$$

The general formula is

$$1^3 + 2^3 + 3^3 + \cdots + n^3 = \left(\frac{n(n+1)}{2}\right)^2.$$

For $n = 6$, since $6^3 = 12 \times 18 = 216$, we obtain $225 + 216 = 441 = 21^2$, or we use the formula $\left(\frac{6 \times 7}{2}\right)^2 = 21^2$.

    Note that

$$(1 + 2 + \cdots + n)^2 = 1^3 + 2^3 + 3^3 + \cdots + n^3.$$

The Product of the Ten Smallest Primes

You can do that by now. We want

$$2 \times 3 \times 5 \times 7 \times 11 \times 13 \times 17 \times 19 \times 23 \times 29.$$

Let's divide the problem into small parts:

$$2 \times 3 \times 5,$$
$$7 \times 11 \times 13,$$
$$17 \times 19,$$
$$23 \times 29.$$

The first product is within the table, $2 \times 3 \times 5 = 30$. Let's apply our methods for the next product. First, we remember $7 \times 13 = 91$. Then we compute $91 \times 11 = 901 + 100 = 1001$.

    For the next case, we remember $17 \times 19 = 323$ and $23 \times 29 = 667$. So, without much trouble we got

$$2 \times 3 \times 5 = 30,$$
$$7 \times 11 \times 13 = 1001,$$
$$17 \times 19 = 323,$$
$$23 \times 29 = 667.$$

Now all we need is to find

$$30 \times 1001 \times 323 \times 667.$$

You may be able to remember these four numbers. Now the last one is approximately two-thirds of 1000. This is actually convenient:

$$30 \times 667 = 30 \times \left(\frac{2}{3} \times 1000 + \frac{1}{3}\right) = 20 \times 1000 + 10 = 20010.$$

The product of this one with 1001 is easy enough to do in your head:

$$1001 \times 20010 = 20030010.$$

Now we multiply this by 323, which is particularly easy:

$$323 \times 20030010 = 6469693230.$$

The trouble is to give it a name: we get six billions, four hundred and sixty-nine millions, six hundred and ninety-three thousands and two hundred and thirty.

Divisibility with a Prime

Divisions by two, three, and five are easy enough just looking at the last digit or taking the sum of the digits. The first problem is dividing by seven.

The theorem says that if the number is written as $a \times 10 + b$ (with b as the last digit), then it is divisible by seven exactly when the number $a - 2 \times b$ is divisible by seven.

The trick works for all primes. We look for the smallest multiple of the prime that differs by only one from a multiple of ten:

$$3 \times 7 = 21,$$
$$1 \times 11 = 11,$$
$$3 \times 13 = 39,$$
$$3 \times 17 = 51,$$
$$1 \times 19 = 19,$$
$$3 \times 23 = 69,$$
$$1 \times 29 = 29.$$

Now we add or subtract this multiple of the prime, a process that does not alter the possibility of divisibility by the prime:

$$10a + b - 21b = 10a - 20b = 10(a - 2b),$$
$$10 + b - 11b = 10a - 10b = 10(a - b),$$

$$10a + b + 39b = 10a + 40b = 10(a + 4b),$$
$$10a + b - 51b = 10a - 50b = 10(a - 5b),$$
$$10a + b + 19b = 10a + 20b = 10(a + 2b),$$
$$10a + b + 69b = 10a + 70b = 10(a + 7b),$$
$$10a + b + 29b = 10a + 30b = 10(a + 3b).$$

For example,

7 :	(-2)	1001	$100 - 2 = 98$	$9 - 16 = -7$
11 :	(-1)	1001	$100 - 1 = 99$	$9 - 9 = 0$
13 :	$(+4)$	1001	$100 + 4 = 104$	$10 + 16 = 26$
17 :	(-5)	323	$32 - 15 = 17$	
19 :	$(+2)$	323	$32 + 6 = 38$	$3 + 16 = 19$
23 :	$(+7)$	667	$66 + 49 = 115$	$11 + 35 = 46$
29 :	$(+3)$	667	$66 + 21 = 87$	$8 + 21 = 29$

For the number 11, this leads to the general rule that 11 divides a number if it divides the alternating sum of the digits, i.e., the sum of every second digit minus the sum of the rest.

The Josephus Problem

It is told that after the Romans conquered Jerusalem in the year 70 A.D., the Jewish writer Josephus, with 40 other Jews, hid in a cellar. They agreed to commit suicide to escape capture by the Romans. Josephus did not really agree, so he suggested the following procedure: they should count 1–2, 1–2, ..., and each time someone said "2" he was decapitated. So they lined up in a circle and the leader started to count. The problem for Josephus was to position himself so that he would be the last to count, so that he could change his mind.

Now you may learn a piece of mathematics. The question is, in which cases is the problem almost trivial? The answer is that for any power of two, it is advantageous to be number one, since after one round you are again number one, and the number of those who remain is half of a power of two, which is again a power of two.

But seeing this, it is easy to solve the general problem: You want to be the first to count when the number of people is reduced to a power of two. In our case of 41, the power is 32, so you must be the one to count after the suicide of nine others. So you want to be number 19.

Spherical Algebra

Istvan Lenart

In this article, we present a binary operation that starts from perpendicular lines on the sphere and concludes with basic elements that consist of a pair of opposite points and their equator.

The Simplest Way of Introducing the Operation

Given two different straight lines on the plane, how many perpendicular straight lines do they have in common? None, if they are intersecting; infinitely many, if they are parallel.

Given two different great circles on the sphere, how many perpendicular great circles do they have in common? Only one, the equator of their two opposite points of intersection.

In other words, we choose two elements in a given order from the set of great circles of a sphere, and find that these two elements determine a great circle, that is, an element of this same set.

This is a binary operation in the set of great circles, just as addition or multiplication is a binary operation in the set of natural or real numbers.

Let's denote this operation by $x \sim y = z$, pronounced "x bang y equals z." By this strange wording, we caution the students

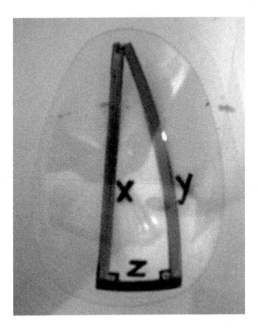

Figure 1. The "bang" operation, $x \sim y = z$.

against confusing this operation with addition or multiplication. (Also, middle- and high-school students like the word itself, and call these lessons "bang-lessons.")

Some Properties of This Operation

Commutativity

It does not matter in Figure 1 whether we draw the great circle on the right first and the one on the left second, or conversely. The equator is the same, if the great circles were the same. So: $x \sim y = y \sim x$.

Cancellation

Consider the operation of addition. *Cancellation* means that, given three numbers x, y, z, for which $x + y = x + z$ holds, then $y = z$ also holds. For multiplication, it is almost the same, except for the $x = 0$ case. So, given three different great circles a, b, and c, for

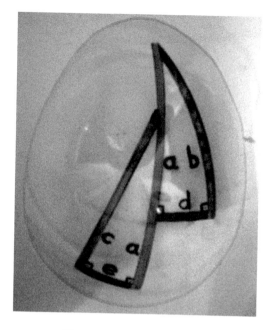

Figure 2. Does a cancel?

which $a \sim b = a \sim c$ holds; is it true that $b = c$ also holds in this case? Can great circle a be canceled on both sides?

First, translate this algebraic condition into a configuration in spherical geometry. Figure 2 depicts $a \sim b$ and $a \sim c$ in the general case, when $a \sim b \neq a \sim c$.

The equation $a \sim b = a \sim c$ means that a, b, c are perpendicular to the same equator; or, in other words, they are *concurrent*. (See Figure 3.)

Cancellation would mean here that concurrency is impossible for three different great circles, because b would inevitably coincide with c. Of course, this is not the case. We can easily draw three concurrent great circles, all of which are different. So cancellation does not work here.

Another Property That Compensates for the Loss of Cancellation

The equation $a \sim b = a \sim c$ and the corresponding configuration of concurrency are so important in algebra and spherical geometry that we try to find some other way of algebraically describing this

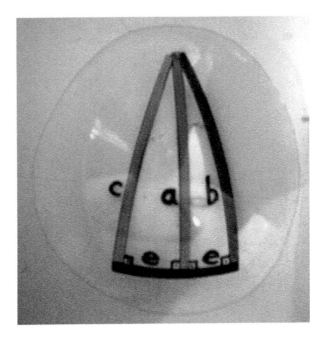

Figure 3. Concurrent great circles, $e = a \sim b = a \sim c$.

geometric configuration. Indeed, the three concurrent great circles suggest the following consequence: $a \sim b = a \sim c$ does not give $b = c$, but it does give $a \sim b = a \sim c = b \sim c$! Let's call this property the *law of symmetrization*.

Law of Symmetrization for Four Elements

Applying the same idea for four concurrent great circles, we find that $a \sim b = c \sim d$ gives $a \sim b = c \sim d = a \sim c = b \sim d = a \sim d = b \sim c$. (See Figure 4.)

Banged by Itself

What can be said about $a \sim a$? If the two great circles coincide, they do not determine a common perpendicular. (See, e.g., Figure 5.) So $a \sim a$ is indefinable, and, for that reason, remains undefined. This means that the "bang" operation is a partial binary operation.

Is this a fiasco of the whole system? Let's refer to the well-known deficiency of division, namely, division by zero. This oper-

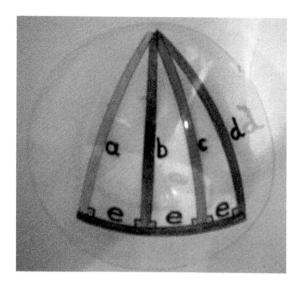

Figure 4. Four concurrent great circles.

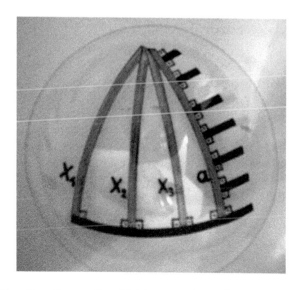

Figure 5. The situation on the right is one potential interpretation of $a \sim a$.

ation is also undefined, but this does not mean the failure of the system as a whole. We add that there are several ways of mapping this geometric system of great circles onto systems involving addition and multiplication among numbers; and these mappings will make the attempted operation of an element banged by itself exactly correspond to division by zero. So the two systems are very close to each other in this kind of lack of definition.

However, this property also means that we have to check whether the two "banged" elements are the same, just as we have to check whether the denominator is zero for each fractional expression.

An Inequality

(It would be logical to call it the *triangle inequality*, but this expression is already used in a different context).

The expression $a \sim b \neq a \sim c$ gives that $a \sim b \neq a \sim c \neq b \sim c \neq a \sim b$. For, $a \sim b \neq a \sim c$ means that a, b, and c are not concurrent, that is, a, b, c form a real spherical triangle. But then $b = c$ cannot be equal to $a \sim b$ or $a \sim c$.

Main Property

What is the result of $(a \sim b) \sim (a \sim c)$?

First, we have to check whether this expression makes sense at all; that is, a, b, c must all be different, and, moreover, $a \sim b \neq a \sim c$. Geometrically, this means that a, b, and c are not concurrent. (On the plane, we would say that three straight lines form a triangle; but three great circles, if not concurrent, form eight different spherical triangles on the sphere.)

In this case, let $a \sim b = p$ and $a \sim c = q$. What is the common perpendicular of p and q? Well, great circle a is perpendicular to p; and great circle a is also perpendicular to q. (See Figure 6.) But, in this case, great circle a is a common perpendicular to p and q; but p and q have only one common perpendicular. So, $p \sim q = a$; therefore, $(a \sim b) \sim (a \sim c) = a$.

General Unit Element

Is there a great circle e for which $x \sim e = x$ holds for all x in the set (except for $e \sim e$, which is undefined)? This would mean that x is perpendicular to itself—but no such great circle exists on the sphere. So, no general unit element exists here.

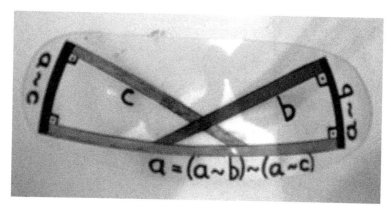

Figure 6. The result of $(a \sim b) \sim (a \sim c)$.

Special Unit Element

As we saw, no general unit element exists in the set of great circles; but perhaps some special unit element does exist. In other words, can we find any element e and another element x for which $e \sim x = x$ holds? No. The same reasoning that we saw above shows that no element of the set has this property, because it would involve self-perpendicularity of a great circle.

Inverse Operation

Given great circles a and b, can we find an element x for which $a \sim x = b$ holds?

Here we have two cases. If a and b are perpendicular to each other, then there are infinitely many solutions for x. If a and b are not perpendicular, no solution of the equation $a \sim x = b$ exists. So, the picture is totally different from the inverse operations that we have been used to. It remains to be seen whether this is a loss or gain.

Perpendiculars in Sets of Three Elements

Given $a \sim b = c$, is it possible that $b \sim c = a$ also holds?

Yes. In this case, the three great circles a, b, and c are such that any two are perpendicular to each other, as shown in Figure 7. Consequently, $c \sim a = b$ must also hold if $a \sim b = c$ and $b \sim c = a$ hold.

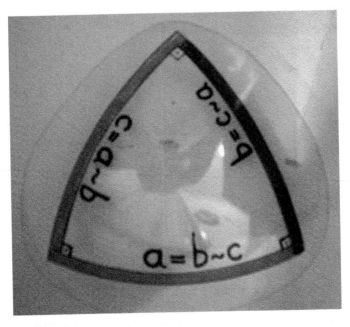

Figure 7. Three great circles such that $a \sim b = c$ and $b \sim c = a$.

It is interesting to try applying this same property for addition or multiplication among numbers. It can be shown that no three different numbers satisfy all the three equations. The proof is easy and short for addition, and also easy, but a bit lengthy, for multiplication.

Question Often Asked by Students

Students often come up with the following question: is it true that $a \sim b = c$ always gives $b \sim c = a$ (and then $c \sim a = b$)? No. Take a spherical triangle with two, not three, right angles. Here $a \sim b = c$, but $b \sim c \neq a$, and $c \sim a \neq b$, either. (We can drop a hint about another algebraic system in which this property is always fulfilled. This is the algebra of *Steiner triplets*.)

Associativity

Given three great circles a, b, and c, does $a \sim (b \sim c) = (a \sim b) \sim c$ always hold, provided that the expression is well defined?

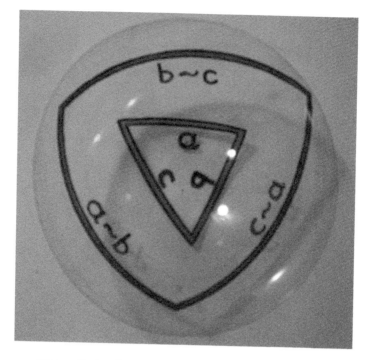

Figure 8. A spherical triangle and its polar triangle.

Take a spherical triangle with sides a, b, and c, and its polar triangle. The sides of the polar triangle will be $b \sim c$, $c \sim a$, and $a \sim b$. (See Figure 8.)

The side that corresponds to side a will be $b \sim c$. So $a \sim (b \sim c)$ is the common perpendicular to these two sides. However, a great circle that is perpendicular to great circle a must pass through the pole point of side a, that is, a vertex of the triangle $b \sim c$, $c \sim a$, $a \sim b$; and a great circle that is perpendicular to great circle $b \sim c$ must pass through a vertex of triangle a, b, c. This means that $a \sim (b \sim c)$ is exactly an altitude of triangle a, b, c—and also an altitude of triangle $b \sim c$, $c \sim a$, $a \sim b$! (See Figure 9.) So an associative element is an altitude of a spherical triangle. Since the three altitudes do not coincide in general, the property of associativity is not valid here.

A consequence: parentheses and brackets will be all-important in this algebra!

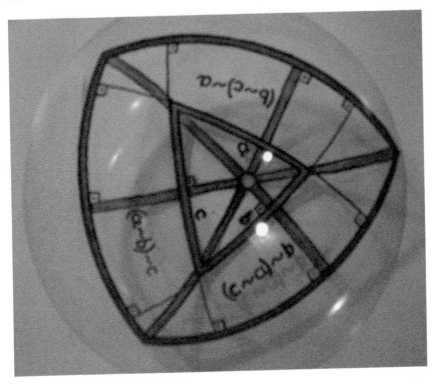

Figure 9. Showing the altitudes of the spherical triangle and its polar triangle from Figure 8.

Existence of the Orthocenter of the Triangle

As we saw above, the associative element $(a \sim b) \sim c$ corresponds to one of the altitudes of the triangle a, b, c. Since the three altitudes are different, the property of associativity is not valid here. But we experience that the three altitudes of a spherical triangle are always concurrent. So, we can state that $(a \sim b) \sim c$, $(b \sim c) \sim a$, and $(c \sim a) \sim b$ are concurrent on the sphere. Or, making use of our notation,

$$[(a \sim b) \sim c] \sim [(b \sim c) \sim a] = [(b \sim c) \sim a] \sim [(c \sim a) \sim b]$$
$$= [(c \sim a) \sim b] \sim [(a \sim b) \sim c].$$

Choosing Our Axioms

We have listed a number of properties. Now the question is: are these properties equally important? Or, to put the point another way, can we pick out a few of these properties, and deduce the other properties, or several of them at least, from these chosen properties, or axioms?

Let's try the following:

> Assume a set S of elements among which a binary operation is defined: $x \sim y = z$. For this operation, two axioms hold:
>
> Axiom 1. $a \sim b = a \sim c$ (commutativity).
>
> Axiom 2. $(a \sim b) \sim (a \sim c) = a$ (main axiom).

From these two axioms, we try to deduce some other properties that we shall refer to as theorems in this system of axioms.

Some Theorems and Their Proofs

Theorem 1. *The bang $a \sim a$ is undefined.*

Proof: We will prove the following: if a and b are different (that is, if the set S has at least two different elements), then $(a \sim b) \sim (a \sim b)$ cannot be unambiguously defined. For $(a \sim b) \sim (a \sim b) = a$ by Axiom 2; but, at the same time, $(a \sim b) \sim (a \sim b) = (b \sim a) \sim (b \sim a) = b$, by Axioms 1 and 2. So $(a \sim b) \sim (a \sim b) = a = b$, which is a contradiction. So $x \sim x$ cannot be properly defined for all x of the set.

□

Theorem 2. *If $a \sim b = a \sim c$, then $a \sim b = a \sim c = b \sim c$.*

Proof: Suppose that a, b, c are all different. We will prove that, if $a \sim b = a \sim c$, then $a \sim b = a \sim c = b \sim c$ also holds. Let's suppose by indirect proof that $a \sim b = a \sim c$, but $a \sim b \neq b \sim c$; then, of course, $a \sim c \neq b \sim c$. We have just proved that $x \sim x$ is undefined. So, if we suppose that $a \sim b \neq b \sim c$, then $a \sim b$ *can* be banged by $b \sim c$, because they are different. The same holds for $a \sim c$ and $b \sim c$, because they are also different. This means that both sides of the equality $a \sim b = a \sim c$ can be "multiplied," or rather banged, by $b \sim c$. We get $(a \sim b) \sim (b \sim c) = (a \sim c) \sim (b \sim c)$. By virtue of the two axioms, we conclude that $b = c$. But we assumed that a, b, c were all different, so this is a contradiction. The only way of solving this contradiction is to suppose that the operation $(a \sim b) \sim (b \sim c)$ cannot be performed, because $a \sim b = b \sim c$; and then, of course, the operation $(a \sim c) \sim (b \sim c)$ cannot be performed either, because $a \sim c = b \sim c$. $\qquad\square$

Theorem 3. *If $a \sim b = c \sim d$, then $a \sim b = c \sim d = a \sim c = b \sim d = a \sim d = b \sim c$.*

Proof: The same proof as above, but for four elements. Suppose that $a \sim b = c \sim d$, but, say, $a \sim c \neq a \sim b$. Banging both sides of the equality $a \sim b = c \sim d$, we get $(a \sim b) \sim (a \sim c) = (c \sim d) \sim (a \sim c)$, and then $a = c$, which is a contradiction. $\qquad\square$

Theorem 4. *If $a \sim b \neq a \sim c$, then $a \sim b \neq a \sim c \neq b \sim c \neq a \sim b$.*

Proof: This is the simple consequence of Theorem 2. If $a \sim b = a \sim c$ gives $a \sim b = a \sim c = b \sim c$, then $a \sim b \neq b \sim c$ gives that $a \sim b$, $a \sim c$, and $b \sim c$ are all different. $\qquad\square$

There is no proof for the special unit element property. Is this property independent of the axioms?

Theorem 5. *If $a \sim b = c$, then $b \sim c = a$.*

Proof: If $a \sim b = c$ and $b \sim c = a$, then $(a \sim b) \sim (b \sim c) = c \sim a$, but $(a \sim b) \sim (b \sim c) = b$, so $a \sim c = b$. $\qquad\square$

Theorem 6. *For three great circles a, b, and c, $a \sim (b \sim c) = (a \sim b) \sim c$.*

Proof: The expression $a \sim (b \sim c) = (a \sim b) \sim c$ gives $a \sim (b \sim c) = (a \sim b) \sim c = a \sim c = (b \sim c) \sim (a \sim b) = b$. So, $a \sim c = b$; but $(a \sim c) \sim b$ is undefined in this case. $\qquad\square$

There is no proof for the existence of the orthocenter of the triangle. Is this property independent of the axioms?

Finite Structures

Our system of axioms is certainly fulfilled by the operation of finding the common perpendicular of two spherical great circles. The set of spherical great circles has infinitely many elements. Can we find a set with only a finite number of elements that also fulfills the two axioms?

Take a set of three elements, 1, 2, and 3. Define the result of the binary operation as the third element of the set. This set and operation can be expressed by a Cayley table:

$\sim$	1	2	3
1	—	3	2
2	3	—	1
3	2	1	—

This set can be proved to fulfill the two axioms. Commutativity is easy to check, and the main axiom can also be checked: for example, $(1 \sim 2) \sim (1 \sim 3) = 3 \sim 2 = 1$.

Now, an interesting trick: let's ruin this table by interchanging the numbers 2 and 3:

$\sim$	1	2	3
1	—	2	3
2	2	—	1
3	3	1	—

Surprisingly, this table will also fulfill the axioms! As before, it is only the main axiom that needs a closer look:

$$(1 \sim 2) \sim (1 \sim 3) = 2 \sim 3 = 1,$$
$$(1 \sim 2) \sim (2 \sim 3) = 2 \sim 1 = 2,$$
$$(1 \sim 3) \sim (2 \sim 3) = 3 \sim 1 = 3.$$

It works! Even more interestingly, these two structures are different, or, in other words, they are not *isomorphic*. This can most easily be seen by observing that no solution of the equation $2 \sim x = 2$ exists in the first case, but there is a solution of this equation in the second case: $x = 1$.

We reach three important conclusions:

1. The two axioms can be fulfilled both in finite and infinite sets.

2. If the set has only three elements, then it can have a general unit element, like the element 1 in the example above. We have to add to Theorem 5 that the set must consist of at least four elements.

3. It was not for nothing that we could not prove the special unit element property. As we see now, special unit elements can exist in such a set.

Here is an example of a set with seven elements:

~	1	2	3	4	5	6	7
1	—	7	1	7	3	3	1
2	7	—	6	7	6	2	2
3	1	6	—	5	6	5	1
4	7	7	5	—	4	5	4
5	3	6	6	4	—	3	4
6	3	2	5	5	3	—	2
7	1	2	1	4	4	2	—

For this set, the Fano, Desargues, Pappos, and Hesse Theorems can be defined and found to be valid. Furthermore, this is the only possible set of seven elements in this algebraic system. For 21 elements, we find two nonisomorphic systems, one of which does not fulfill the Hesse Theorem. For 91 elements, there are four nonisomorphic systems, two of which do not fulfill any of the above-listed theorems.

A Shift in the Geometric Foundation

Until now, our geometric basic element has been the great circle. However, a great circle determines two pole points on the sphere; and conversely, any spherical point determines one equator. So we can choose for the basic element of our theory not a great circle, not a spherical point, not even a pair of opposite points (as in Riemannian geometry), but the unification of a point *and* its opposite point *and* their equator, taking them all together. It can be checked that any two different basic elements of this kind determine a third basic element of the same kind. The geometric operation can be expressed in several ways, which all mean the same from our algebraic point of view:

1. the great circle through two not-opposite points,

2. the common perpendicular of two differentgreat circles,

3. the perpendicular dropped from a point to a great circle that is different from the point's equator,

4. the pair of opposite points at a distance of 90 degrees from two not-opposite points.

On the basis of this new standpoint, we can check again all the properties. For example, the geometric content of the main axiom can also be formulated the following way: Take the three vertices of a triangle. Starting from a vertex, draw two sides of the triangle. The point of intersection of the two sides gives the starting vertex back again.

A Way of Constructing Such a Set on the Elements of a Field

Assume a field F with two operations, addition and multiplication. Consider all the ordered triplets of this set. Exclude $(0,0,0)$. Any two triplets (a,b,c) and (d,e,f) determine the same element if $a = kd$, $b = ke$, and $c = kf$. Now the operation between the triplets (m,n,o) and (p,q,r) is as follows:

$$mx + ny + oz = 0,$$
$$px + qy + rz = 0.$$

Mathematical Idol

Colm Mulcahy

You stand before an audience, a highly-trained assistant by your side. "I'm going to perform a mathematical experiment, with lots of input from you, and a little help from my assistant here. In a moment, while she is out of the room, you'll rank eight famous contestants for the grand title of *Mathematical Idol*, starting by picking the three finalists. Then one of you will ask my assistant to rejoin us, and I'll tell her the second and third place runners-up, no more. She will then try to figure out who got first place. If she succeeds, I'll then tell her two more of the remaining runners-up, which you'll already have ranked. She'll attempt to fill in the three missing ones, in the correct order!" Of course, your two-stage experiment is successful.

Here's the set-up in more detail. Introduce your mathematically savvy assistant and then have her leave the room. Hand out eight index cards, each bearing the name of a famous mathematician on one side. The goal is to have the audience assist with the determination of a full ranking of these eight, with the mathematician in first place being declared Mathematical Idol. Have the cards turned face-down and mixed. Ask for three of them back. As you take them remark, "By giving me these three finalists, you are eliminating five; we'll come back to those in a moment."

Line up those cards face-down on the table, pointing to the last one in the row, declaring, "That's the overall winner, Mathematical Idol! We'll find out who it is presently."

Continue, "A moment ago, you eliminated five of the eight contestants, but they too deserve to be ranked. Please hand me those cards one by one, in whatever order you wish, and I'll extend this row on the table yielding a full ranking from eighth to first place." As the cards are handed to you, supplement the existing row, placing these five cards face-up, starting at the end where the third place winner is. Finally, turn these cards face-down also, so that none of the eight cards are identifiable.

"You, the audience, have acted as the judges here, first eliminating five of the eight contestants, then ranking the three finalists, before finally ranking the five eliminated mathematicians. So many different possibilities! Can somebody please step outside and bring my assistant back in?"

As promised, you first reveal the runners-up among the three finalists to your assistant—literally turning over those two cards—and sure enough, she soon names the overall winner. That card is turned over to confirm the accuracy of her prediction.

Next, as the applause dies down, you reveal two of the remaining five runners-up to your assistant. She promptly names the hidden ones—*in the correct order.*

The Mathematics Behind the Drama

There are two distinct mathematical principles at work here. The first one allows your accomplice to determine the overall winner. If you wish, you and your accomplice can perform only this part of the trick and ignore the second part entirely.

Also, this first part can be adapted as a telephone trick.

The First Principle

Here's is a curious but little known fact:

> If three items are chosen from eight, then two of the three can be arranged in a row to predict the identity of the third.

It's a fun and easy exercise to figure out ways to do this if the choices are made from six items, and such strategies can be modified to work for seven, but if the selections are made from eight, it's a bit trickier. There's no hope for more than eight (can you see why?).

This is a special case of a more general result that is often associated with a two-person card trick. What is crucial is the ranking of those three finalists, which *you* determine, after performing some mental calculations explained below. From that ranking alone, your assistant can indeed figure out who the overall winner must be. It's a case of mathematical, not physical or verbal, communication!

(For a more magical performance, one could utilize the "magician's choice" method to force this part of the proceedings, thus giving the illusion that you also leave the ordering of those three finalists up to chance.)

Let's take a closer look at the numbers. You start with one of $\frac{8!}{3!5!} = 56$ selections of three objects from eight, and you will then decide on just one of the $3! = 6$ possible ways to rank them. Meanwhile, your assistant learns of one of the $8 \times 7 = 56$ possible runners-up arrangements (for the third and second places). She is expected to figure out the overall winner from that information alone, and she succeeds, because $56 \leq 56$ *and* $56 \geq 56$!

The trick is to come up with a logical one-to-one correspondence between the 56 (ordered) runners-up arrangements and the 56 (unordered) sets of three finalists, so that both of you can perform this live in real time. There are many ways to do this, but the one we explain here readily generalizes to larger numbers. We'll implement (the simplest nontrivial case of) an algorithm due to Elwyn Berlekamp, which has appeared in print [2, 3] several times in recent years in connection with the optimal version of a famous two-person card trick.

The popular incarnation of the general result says: for a deck of (up to) $n! + n - 1$ cards, if we chose any n at random, it is possible to arrange some $n - 1$ of them in a row in such a way as to indicate the identity of the remaining one. The famous card trick is the case $n = 5$, which is easy enough for a standard 52-card deck, but tricky for one of size 124.

Our focus here is on the case $n = 3$. We follow the treatment found in a delightful online Maple worksheet by John Cosgrave, of St. Patrick's College, Dublin [1].

First we identify Archimedes, Euclid, Euler, Gauss, Hilbert, Newton, Pythagoras, and Riemann with 0, 1, 2, 3, 4, 5, 6, and 7, respectively. Suppose that $c_0 < c_1 < c_2$ are chosen from $0, 1, 2, \ldots, 7$. The *winning label* $w = c_0 + c_1 + c_2 \pmod 3$ determines the *winner* c_w (Mathematical Idol). You convey this fact to your assistant by revealing the remaining two c_j's in the appropriate order. Note that $c_w \geq w$; in fact, considering the cases $w = 0, 1, 2$, in turn, we see that

> $c_w - w \in \{0, 1, 2, \ldots, 5\}$ gives the position of c_w if those c_j's that are smaller than it are eliminated, so that its position is "bumped back" a value or two if necessary.

Next, you compute the sum s of the two runners-up. Since $s = w - c_w \pmod 3$, then the nonnegative number $c_w - w$ is congruent to $-s \pmod 3$. Your assistant can figure out $c_w - w \pmod 3$ for herself, as she knows the two runners-up, and hence s. By observing the order of these finalists, she can deduce the exact value of $c_w - w$, and hence, as we will see, the winner c_w. Here's how:

By calculating $-s \pmod 3$, your assistant narrows down the value of $c_w - w$ to one of two possibilities, differing by 3. That's where the order of the two runners-up comes in: you use ascending order to indicate the first possibility (0, 1, or 2), and descending for the second (3, 4, or 5). Hence your assistant can determine $c_w - w$ exactly. All that remains for her to do is figure out how much (0, 1, or 2) that value should be bumped up by to correctly identify c_w, namely Mathematical Idol.

For example, suppose that the three finalists are 0, 2, and 6. Since $w = 0 + 2 + 6 = 2 \pmod 3$, then the Mathematical Idol is $c_2 = 6$. Note that $c_w - w = 6 - 2 = 4$; this is what must be coded by the two runners-up 0 and 2. The sum of these is $s = 0 + 2 = 2 \pmod 3$, and $c_w - w = -s = 1 \pmod 3$ as expected. The negative of the sum s reveals $1 \pmod 3$ to your assistant, and if the two finalists in question are displayed in descending order 2, 0, then that communicates the fact that $c_w - w = 1 + 3 = 4$ (as opposed to 1). So your assistant can infer that $c_w \geq 4$, and of the three possibilities available, only $c_w = 6$ works. So $w = 2$, and Mathematical Idol must be $c_2 = 6$ (namely, Riemann).

Let's look at a second example, this time from your assistant's perspective only. Suppose that she sees runners-up 1 and 6, in that order. She computes $s = 1 + 6 = 1 \pmod 3$, deducing that

$c_w - w = -s = 2 \pmod 3$. So she knows that either $c_w - w = 2$ or 5. Since the two runners-up are in ascending order, she infers that $c_w - w = 2$. Hence, $c_w \geq 2$, and of the three possibilities available, only $c_w = 3$ works. So $w = 1$, and Mathematical Idol must be $c_1 = 3$ (namely, Gauss).

Live performance may be facilitated by customized cheat sheets, one for you (Table 1) and another for your assistant (Table 2). Yours is alphabetized (in the left column) by finalists, showing both the winner (in italics) and (in the right column) the order in which the runners-up are shown to your assistant. Your assistant's is alphabetized (in the right column) by the runners-up (in order), showing (in the right column) all three finalists, with the winner in italics. Your assistant's is just the same as yours except that the rows are reordered for ease of lookup: it's alphabetized by the entries in the second column, as opposed to the first.

The Second Principle

Once the third and second place runners-up have enabled your accomplice to name the overall winner, she, like you, knows the identities of the remaining five contestants. They have earlier been laid out in an order that you saw but could not control. Yet, because of the following result, it's possible for you to show her just two of them and surreptitiously communicate the identity *and order* of the remaining three.

> In any list of five distinct numbers, there is either a rising subsequence of length three or a falling subsequence of length three.

Given that, you proceed as follows. If there is a rising subsequence of length three, then you turn over the other two cards from left to right. If there is a falling subsequence of length three, then you turn over the other two cards from right to left. In both cases your assistant knows what three are missing and deduces their order (rising or falling) based on the way in which you revealed the two cards. (It's not so obvious how to pull this part off over the telephone!)

Here's a proof of the subsequence fact. Suppose we have five numbers A, B, C, D, and E, with A < B (the argument when A > B is similar). Assume that there is not a rising subsequence (from left to right) of length three; we'll show that there is a falling one

Finalists (winner in *italics*)	Runners-up in order
Archimedes, Euclid, Euler	Euclid, Euler
Archimedes, *Euclid*, Gauss	Archimedes, Gauss
Archimedes, Euclid, *Hilbert*	Archimedes, Euclid
Archimedes, Euclid, Newton	Euclid, Newton
Archimedes, *Euclid*, Pythagoras	Archimedes, Pythagoras
Archimedes, Euclid, *Riemann*	Euclid, Archimedes
Archimedes, Euler, *Gauss*	Archimedes, Euler
Archimedes, Euler, Hilbert	Euler, Hilbert
Archimedes, *Euler*, Newton	Archimedes, Newton
Archimedes, Euler, *Pythagoras*	Euler, Archimedes
Archimedes, Euler, Riemann	Euler, Riemann
Archimedes, *Gauss*, Hilbert	Archimedes, Hilbert
Archimedes, Gauss, *Newton*	Gauss, Archimedes
Archimedes, Gauss, Pythagoras	Gauss, Pythagoras
Archimedes, *Gauss*, Riemann	Archimedes, Riemann
Archimedes, Hilbert, Newton	Hilbert, Newton
Archimedes, *Hilbert*, Pythagoras	Pythagoras, Archimedes
Archimedes, Hilbert, *Riemann*	Hilbert, Archimedes
Archimedes, Newton, *Pythagoras*	Newton, Archimedes
Archimedes, Newton, Riemann	Newton, Riemann
Archimedes, *Pythagoras*, Riemann	Riemann, Archimedes
Euclid, Euler, Gauss	Euler, Gauss
Euclid, *Euler*, Hilbert	Euclid, Hilbert
Euclid, Euler, *Newton*	Euler, Euclid
Euclid, Euler, Pythagoras	Euler, Pythagoras
Euclid, *Euler*, Riemann	Euclid, Riemann
Euclid, Gauss, *Hilbert*	Euclid, Gauss
Euclid, Gauss, Newton	Gauss, Newton
Euclid, *Gauss*, Pythagoras	Euclid, Pythagoras
Euclid, Gauss, *Riemann*	Gauss, Euclid
Euclid, *Hilbert*, Newton	Newton, Euclid
Euclid, Hilbert, *Pythagoras*	Hilbert, Euclid
Euclid, Hilbert, Riemann	Hilbert, Riemann
Euclid, Newton, Pythagoras	Newton, Pythagoras
Euclid, *Newton*, Riemann	Riemann, Euclid
Euclid, Pythagoras, *Riemann*	Pythagoras, Euclid
Euler, Gauss, Hilbert	Gauss, Hilbert
Euler, *Gauss*, Newton	Euler, Newton
Euler, Gauss, *Pythagoras*	Gauss, Euler
Euler, Gauss, Riemann	Gauss, Riemann
Euler, Hilbert, *Newton*	Hilbert, Euler
Euler, Hilbert, Pythagoras	Hilbert, Pythagoras
Euler, *Hilbert*, Riemann	Riemann, Euler
Euler, *Newton*, Pythagoras	Pythagoras, Euler
Euler, Newton, *Riemann*	Newton, Euler
Euler, Pythagoras, Riemann	Pythagoras, Riemann
Gauss, Hilbert, Newton	Newton, Hilbert
Gauss, *Hilbert*, Pythagoras	Pythagoras, Gauss
Gauss, Hilbert, *Riemann*	Hilbert, Gauss
Gauss, Newton, *Pythagoras*	Newton, Gauss
Gauss, Newton, Riemann	Riemann, Newton
Gauss, *Pythagoras*, Riemann	Riemann, Gauss
Hilbert, Newton, Pythagoras	Pythagoras, Newton
Hilbert, *Newton*, Riemann	Riemann, Hilbert
Hilbert, Pythagoras, *Riemann*	Pythagoras, Hilbert
Newton, Pythagoras, Riemann	Riemann, Pythagoras

Table 1. Performer's cheat sheet.

Finalists (winner in *italics*)	Runners-up in order
Archimedes, Euclid, *Hilbert*	Archimedes, Euclid
Archimedes, Euler, *Gauss*	Archimedes, Euler
Archimedes, *Euclid*, Gauss	Archimedes, Gauss
Archimedes, *Gauss*, Hilbert	Archimedes, Hilbert
Archimedes, *Euler*, Newton	Archimedes, Newton
Archimedes, *Euclid*, Pythagoras	Archimedes, Pythagoras
Archimedes, *Gauss*, Riemann	Archimedes, Riemann
Archimedes, Euclid, *Riemann*	Euclid, Archimedes
Archimedes, Euclid, Euler	Euclid, Euler
Euclid, Gauss, *Hilbert*	Euclid, Gauss
Euclid, *Euler*, Hilbert	Euclid, Hilbert
Archimedes, Euclid, Newton	Euclid, Newton
Euclid, *Gauss*, Pythagoras	Euclid, Pythagoras
Euclid, *Euler*, Riemann	Euclid, Riemann
Archimedes, Euler, *Pythagoras*	Euler, Archimedes
Euclid, Euler, *Newton*	Euler, Euclid
Euclid, Euler, Gauss	Euler, Gauss
Archimedes, Euler, Hilbert	Euler, Hilbert
Euler, *Gauss*, Newton	Euler, Newton
Euclid, Euler, Pythagoras	Euler, Pythagoras
Archimedes, Euler, Riemann	Euler, Riemann
Archimedes, Gauss, *Newton*	Gauss, Archimedes
Euclid, Gauss, *Riemann*	Gauss, Euclid
Euler, Gauss, *Pythagoras*	Gauss, Euler
Euler, Gauss, Hilbert	Gauss, Hilbert
Euclid, Gauss, Newton	Gauss, Newton
Archimedes, Gauss, Pythagoras	Gauss, Pythagoras
Euler, Gauss, Riemann	Gauss, Riemann
Archimedes, Hilbert, *Riemann*	Hilbert, Archimedes
Euclid, Hilbert, *Pythagoras*	Hilbert, Euclid
Euler, Hilbert, *Newton*	Hilbert, Euler
Gauss, Hilbert, *Riemann*	Hilbert, Gauss
Archimedes, Hilbert, Newton	Hilbert, Newton
Euler, Hilbert, Pythagoras	Hilbert, Pythagoras
Euclid, Hilbert, Riemann	Hilbert, Riemann
Archimedes, Newton, *Pythagoras*	Newton, Archimedes
Euclid, *Hilbert*, Newton	Newton, Euclid
Euler, Newton, *Riemann*	Newton, Euler
Gauss, Newton, *Pythagoras*	Newton, Gauss
Gauss, Hilbert, Newton	Newton, Hilbert
Euclid, Newton, Pythagoras	Newton, Pythagoras
Archimedes, Newton, Riemann	Newton, Riemann
Archimedes, *Hilbert*, Pythagoras	Pythagoras, Archimedes
Euclid, Pythagoras, *Riemann*	Pythagoras, Euclid
Euler, *Newton*, Pythagoras	Pythagoras, Euler
Gauss, *Hilbert*, Pythagoras	Pythagoras, Gauss
Hilbert, Pythagoras, *Riemann*	Pythagoras, Hilbert
Hilbert, Newton, Pythagoras	Pythagoras, Newton
Euler, Pythagoras, Riemann	Pythagoras, Riemann
Archimedes, *Pythagoras*, Riemann	Riemann, Archimedes
Euclid, *Newton*, Riemann	Riemann, Euclid
Euler, *Hilbert*, Riemann	Riemann, Euler
Gauss, *Pythagoras*, Riemann	Riemann, Gauss
Hilbert, *Newton*, Riemann	Riemann, Hilbert
Gauss, Newton, Riemann	Riemann, Newton
Newton, Pythagoras, Riemann	Riemann, Pythagoras

Table 2. Assistant's cheat sheet.

instead. We must have each of C, D, and E less than B, as otherwise there would be a rising subsequence of length three starting with A, B. If $C > D$, then B, C, D would be a falling subsequence of length three, whereas if $D > E$, then B, D, E would be a falling subsequence of length three; in both cases we are done. Actually, one of those subcases must occur, for otherwise both $C < D$ and $D < E$, and hence C, D, E would be a rising subsequence of length three, contrary to our assumption.

This is the first nontrivial case of an old result from the 1930s, due to Hungarians Paul Erdős and Gyorgy Szekeres. It asserts that

> Given a list of $k^2 + 1$ distinct numbers, there is either an increasing subsequence of size $k + 1$ or a decreasing subsequence of size $k + 1$.

We just encountered the special case $k = 3$ and $n = 5$.

Performance Time

You stand before an audience, an assistant by your side. You do all the talking at first. "Welcome to Mathematical Idol! Thanks for coming. Hundreds of top mathematicians from down through the ages have worked hard to be with us here today, proving theorems, formulating conjectures and providing inspiration for countless students. In a moment, *you* will select the overall winner, Mathematical Idol! First prize is an indefinite publishing contract with a transfinite number of the profession's leading journals and an *e*-course dinner with Paul Erdős at the *Theorems from Coffee Café* at Hilbert's Hotel." Pause for applause. "After numerous preliminary Gaussian elimination rounds, we are down to just eight contestants: Archimedes, Euclid, Euler, Gauss, Hilbert, Newton, Pythagoras, and Riemann. Please give these guys a big welcome!"

Display eight index cards, each with the name of one of these mathematicians. You might wish to include biographical snippits, some of which you can read out in due course, such as:

> Some would argue that Carl Friedrich Gauss was the greatest talent to ever emerge in his profession. Carl's string of top hits range from "Quadratic Reciprocity Love," "The Fundamental Theorem of Algebra," and "Hypergeometric Funky," to "Have You Seen My Regular 17-gon, Baby?"

or

> Euclid's *Elements* is the best selling release of all time, having been in the charts now for well over 2000 years. Euclid may have sold more than Frank Sinatra, Elvis Presley, the Beatles, and U2 combined—he's certainly lasted longer. Watch out for the new Elvis Costello remix of "Thirteen Steps Lead to a Proof."

"Normally, we'd have the general public, voting by phone, and our panel of impartial, expert judges, to assist with shortening this down to three names, for the exciting finale. But today, we are going to let one of *you* determine who the three finalists are." Hand out all eight index cards, and request that they be thoroughly mixed. "In a moment, we are going to do a mindreading experiment, so I'd like to ask my assistant here to leave us for a while." Motion to your assistant. "Please step outside for a moment, somebody will call you back in when we are ready."

As soon as your assistant is safely out of the room and out of earshot, ask for any three of the index cards back. Glance at them and pass them from hand to hand, commenting, "You the audience have selected three finalists. I will play the role of the panel of expert judges, and rank these three, into third place, second place, and overall winner." Place the three index cards, face-down, in a row on the table, then point to the first one and say, "That is the winner—Mathematical Idol! Beside it are the two runners-up. Before we ask my assistant back to the room, please rank the other five contestants too."

Have the other five cards appended to the existing row face-up, in any order, so that all eight cards are now on display. Look at these carefully and ask if anyone wishes to alter the order of the five in positions eight to four. Finally, turn these face-down too, and have your assistant fetched.

As soon as she returns, say, "After you left us, we went through a scrupulously fair procedure to rank the eight contestants. The results are displayed here, face-down. Here's the card that identifies the third place runner-up. Please read out that name." Hand over the corresponding card, and the name is read out. Repeat for the mathematician in second place.

"So the big question is, who is Mathematical Idol? It could be any of the other six mathematicians from the list of eight contestants we all saw earlier." Needless to say, your assistant soon announces a name. Have somebody turn over the index card on the table and confirm the correctness of the identification. "Ladies, and gentleman, we have a winner: Mathematical Idol!"

Next, turn over two of the remaining five cards, as indicated before. Once more, your assistant announces the names of the other three in order, and is promptly confirmed to be correct this time too.

Bibliography

[1] John Cosgrave. "Challenging Mathematical Puzzles and Problems." Available at http://services.spd.dcu.ie/johnbcos/challenging. htm, 2007.

[2] Michael Kleber. "The Best Card Trick." *Mathematical Intelligencer* 24:1 (2002), 9–11.

[3] Shai Simonson and Tara Holm. "Using a Card Trick to Teach Discrete Mathematics." *PRIMUS* XIII:3 (2003), 248–269.

The Elevator Problem

David Rhee and Jerry Lo

In the building shown in Figure 1, there are four elevators—A, B, C, and D—each stopping on three of the five floors. Note that an elevator is not required to stop on consecutive floors, or on the first floor.

The elevator system in this building has a very nice property. One can go from any floor to any other floor without having to change elevators. The first and second floors are linked by elevators A, B, and C, and they are linked to the third, fourth and fifth floors by elevators A, B, and C, respectively. The third, fourth and fifth floors are linked by elevator D. A building with such an elevator system is said to be *convenient*. It is said to be *perfect* if each pair of floors is linked by exactly one elevator.

More generally, let there be m elevators in a building, each stopping on n of the floors. We wish to determine the maximum number $f(m,n)$ of floors this building can have if the elevator system is to be convenient. Our initial example shows that $f(4,3) \geq 5$. It will be shown later that $f(4,3) = 5$.

We first prove several useful preliminary results.

Lemma 1. $f(m+1,n) \geq f(m,n)$.

Proof: Having an extra elevator never hurts, though it may not help. $\square$

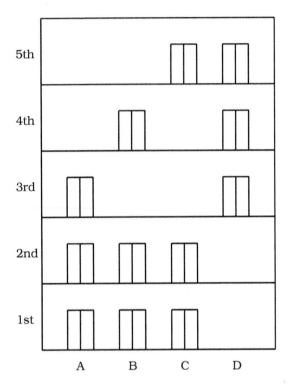

Figure 1. The elevator system.

Lemma 2. $f(m, n+1) \geq f(m, n) + 1$.

Proof: The extra stop for each elevator can all be on a new floor. □

Lemma 3. $f(m, kn) \geq kf(m, n)$.

Proof: Pile k copies of a convenient building with $f(m, n)$ floors on top of one another to form a building with $kf(m, n)$ floors and connect the corresponding elevators in each copy so that each stops on kn floors. The same elevator that links the ith and jth floors in each copy will link the ith floor of any copy to the jth floor of any other copy. Thus the new building is convenient, and we have $f(m, kn) \geq kf(m, n)$.

Keeping n Constant

We first study the function $f(m,n)$ by keeping n fixed.

Theorem 1. $f(m,1) = 1$.

Proof: The building can certainly have one floor. If it has two floors, no elevator will stop on both. □

Theorem 2. $f(m,2) = k$ *where* $\binom{k}{2} \le m < \binom{k+1}{2}$.

Proof: Each elevator can link $\binom{2}{2} = 1$ pair of floors. In a building with k floors, there are $\binom{k}{2}$ pairs of floors. Since $\binom{k}{2} \le m$, m elevators are sufficient. In a building with $k+1$ floors, there are $\binom{k+1}{2}$ pairs of floors. Since $\binom{k+1}{2} > m$, m elevators are not sufficient. □

For $n = 3$, we consider $f(m,3)$ for small values of m. It is easy to prove that $f(1,3) = f(2,3) = 3$, $f(3,3) = 4$, $f(4,3) = f(5,3) = 5$ and $f(6,3) = 6$. The next four values are identical.

Case Study 1. $f(7,3) = f(8,3) = f(9,3) = f(10,3) = 7$.

Proof: Let the first elevator stop on floors 1, 2, and 3; the second on 1, 4, and 5; the third on 1, 6, and 7; the fourth on 2, 4, and 6; the fifth on 2, 5, and 7; the sixth on 3, 4, and 7; and the seventh on 3, 5, and 6. This shows that $f(7,3) \ge 7$. We claim that $f(10,3) \le 7$. The total number of stops is 30. If each floor is served by at least four elevators, then the number of floors is at most seven. If some floor is served by at most three elevators, it can be linked to at most six other floors. Counting this floor, the building can have at most seven floors. Since $f(7,3) \le \cdots \le f(10,3)$ by Lemma 1, we have $f(7,3) = f(8,3) = f(9,3) = f(10,3) = 7$. □

After $f(11,3) = 8$, we have another block of four identical values.

Case Study 2. $f(12,3) = f(13,3) = f(14,3) = f(15,3) = f(16,3) = 9$.

Proof: Let the first elevator stops on floors 1, 2, and 3; the second on 1, 4, and 7; the third on 1, ,5 and 9; the fourth on 1, 6, and 8; the fifth on 2, 4, and 9; the sixth on 2, 5, and 8; the seventh on 2, 6, and 7; the eighth on 3, 4, and 8; the ninth on 3, 5, and 7; the tenth on 3, 6, and 9; the eleventh on 4, 5, and 6; and the twelfth on 7, 8, and 9. This shows that $f(12,3) \ge 9$. We claim that

$f(16,3) \leq 9$. The total number of stops is 48. If each floor is served by at least five elevators, then the number of floors is at most nine. If some floor is served by at most four elevators, it can be linked to at most eight other floors. Counting this floor, the building can have at most nine floors. Since $f(12,3) \leq \cdots \leq f(16,3)$ by Lemma 1, we have $f(12,3) = f(13,3) = f(14,3) = f(15,3) = f(16,3) = 9$.   □

At this point, it is evident that $f(m,3)$ behaves rather erratically. We abandon this approach.

Keeping m Constant

We now study the function $f(m,n)$ by keeping m fixed.

Theorem 3. $f(1,n) = n$.

Proof: The building can certainly have n floors. If it has more, the elevator will not stop on some floor. No elevator will stop on both this floor and any other floor.   □

Theorem 4. $f(2,n) = n$.

Proof: By Lemma 1, $f(2,n) \geq f(1,n) = n$. If the building has more floors, each elevator will not stop on some floor. If they skip different floors, no elevator will stop on both. If they skip the same floor, no elevator will stop on both this floor and any other floor.   □

Theorem 5. $f(3,2k) = 3k$ *and* $f(3,2k+1) = 3k+1$.

Proof: By Theorem 2, $f(3,2) = 3$. By Lemma 3, $f(3,2k) \geq 3k$. The total number of stops is $6k$. If each floor is served by at least two elevators, then the number of floors is at most $3k$. If some floor is served by at most one elevator, it can be linked to at most $2k-1$ other floors. Counting this floor, the building can have at most $2k$ floors. It follows that $f(3,2k) = 3k$. By Lemma 2, $f(3,2k+1) \geq f(3,2k)+1 = 3k+1$. The total number of stops is $6k+3$. If each floor is served by at least two elevators, then the number of floors is at most $3k+1$. If some floor is served by at most one elevator, it can be linked to at most $2k$ other floors. Counting this floor, the building can have at most $2k+1$ floors. It follows that $f(3,2k+1) = 3k+1$.□

We are unable to get any general results for $m = 4$ or 5, but we are convinced that we are on the right track, largely because of Lemma 3. The difficulty here is the absence of perfect buildings for small values of n. The next perfect building occurs at $m = 6$.

Theorem 6. $f(6, 2k) = 4k$ and $f(6, 2k + 1) = 4k + 2$.

Proof: By Theorem 2, $f(6, 2) = 4$. By Lemma 3, $f(6, 2k) \geq 4k$. The total number of stops is $12k$. If each floor is served by at least three elevators, then the number of floors is at most $4k$. If some floor is served by at most two elevators, it can be linked to at most $4k - 2$ other floors. Counting this floor, the building can have at most $4k - 1$ floors. It follows that $f(6, 2k) = 4k$. To prove that $f(6, 2k + 1) \leq 4k + 2$, observe that the total number of stops is $12k + 6$. If each floor is served by at least three elevators, then the number of floors is at most $4k + 2$. If some floor is served by at most two elevators, it can be linked to at most $4k$ other floors. Counting this floor, the building can have at most $4k + 1$ floors. We now give a general construction to show that $f(6, 2k + 1) \geq 4k + 2$. Let the floors be $a_1, a_2, \ldots, a_k, b_1, b_2, \ldots, b_k, c_1, c_2, \ldots, c_k, d_1, d_2, \ldots, d_k, e, f$. Let the first elevator stop at all the a's and b's, the second at all the a's and c's, the third at all the a's and d's, the fourth at all the b's and c's, the fifth at all the b's and d's, and the sixth at all the c's and d's. Then these $4k$ floors are all linked. If we add e as the last stop of the first and sixth elevator and f as the last stop of the second and fifth elevator, they are also linked to the other $4k$ floors. However, e and f are not linked. So we replace d_k by f in the sixth elevator. This destroys the links between d_k on the one hand and e and the c's on the other. The remedy is to add e as the last stop of the third elevator and d_k as the last stop of the fourth elevator. It follows that $f(6, 2k + 1) = 4k + 2$. □

The next perfect building where $n = 2$ occurs at $m = 10$, but Lemma 3 is no longer sharp. We can only get one isolated value, namely, $f(10, 4) = 10$. For $k \geq 3$, we have $10k \leq f(10, 2k) \leq 6k - 2$.

So we search for perfect buildings where $n = 3$. Fortunately, there are infinitely many examples here, provided by the so-called Steiner Triple Systems. The first two examples occur at $m = 7$ and 12. However, their values of m grow rapidly wider.

Theorem 7. $f(7, 3k) = 7k$ and $f(7, 3k + 2) = 7k + 4$.

Proof: By Case Study 1, $f(7,3) = 7$. By Lemma 3, $f(7,3k) \geq 7k$. The total number of stops is $21k$. If each floor is served by at least three elevators, then the number of floors is at most $7k$. If some floor is served by at most two elevators, it can be linked to at most $6k - 2$ other floors. Counting this floor, the building can have at most $6k - 1$ floors. It follows that $f(7,3k) = 7k$. To prove that $f(7,3k+2) \leq 7k + 4$, observe that the total number of stops is $21k + 14$. If each floor is served by at least three elevators, then the number of floors is at most $7k + 4$. If some floor is served by at most two elevators, it can be linked to at most $6k + 2$ other floors. Counting this floor, the building can have at most $6k + 3$ floors. We now give a general construction to show that $f(7,3k+2) \geq 7k + 4$. Let the floors be a_1, $a_2, \ldots, a_{k+1}, b_1, b_2, \ldots, b_{k+1}, c_1, c_2, \ldots, c_{k+1}, d_1, d_2, \ldots, d_{k+1}, e_1, e_2,$ $\ldots, e_k, f_1, f_2, \ldots, f_k, g_1, g_2, \ldots, g_k$. Let the first elevator stop at all the a's, b's, and e's; the second at all the a's, c's, and f's; the third at all the a's, d's, and g's; the fourth at all the b's, c's, and g's; the fifth at all the b's, d's, and f's; the sixth at all the c's, d's, and e's; and the seventh at all the e's, f's, and g's. Then all the floors are linked, with two wasted stops in the seventh elevator. It follows that $f(7,3k+2) = 7k + 4$. $\qquad\square$

For $f(7,3k+1)$, the lower bound is $7k + 1$ while the upper bound is $7k + 2$.

Theorem 8. $f(12,3k) = 9k$.

Proof: By Case Study 2, $f(12,3) = 9$. By Lemma 3, $f(12,3k) \geq 9k$. The total number of stops is $36k$. If each floor is served by at least four elevators, then the number of floors is at most $9k$. If some floor is served by at most three elevators, it can be linked to at most $9k - 2$ other floors. Counting this floor, the building can have at most $9k - 1$ floors. It follows that $f(12,3k) = 9k$. $\qquad\square$

For $n \geq 4$, we can look to the so-called Finite Affine Planes and Finite Projective Planes. The first example occurs at $(m,n) = (13,4)$. However, each of the two types of structures provides at most one value of m for each n.

Theorem 9. $f(13,4k) = 13k$.

Proof: Let the first elevator stop on floors 1, 2, 3, and 10; the second on 4, 5, 6, and 10; the third on 7, 8, 9, and 10; the fourth on 1, 5, 9, and 11; the fifth on 2, 6, 7, and 11; the sixth on 3, 4, 8,

m \ n	1	2	3	4	5	6	7	8	9	10	11	12	13	14	15	16
1	**1**	2	3	4	5	6	7	8	9	10	11	12	13	14	15	16
2	1	**2**	3	4	5	6	7	8	9	10	11	12	13	14	15	16
3	1	**3**	4	6	7	9	10	12	13	15	16	18	19	21	22	24
4	1	3	5													
5	1	3	5													
6	1	**4**	6	8	10	12	14	16	18	20	22	24	26	28	30	32
7	1	4	**7**		11	14		18	21		25	28			32	35
8	1	4	7													
9	1	4	7													
10	1	**5**	7	10												
11	1	5	8													
12	1	5	**9**			18			27			36			45	
13	1	5	9	**13**				26				39				52
14	1	5	9													
15	1	**6**	9													
16	1	6	9													

Table 1. Summary of findings for $1 \leq m, n \leq 16$.

and 11; the seventh on 1, 6, 8, and 12; the eighth on 2, 4, 9, and 12; the ninth on 3, 5, 7, and 12; the tenth on 1, 4, 7, and 13; the eleventh on 2, 5, 8, and 13; the twelfth on 3, 6, 9, and 13; and the thirteenth on 10, 11, 12, and 13. This shows that $f(13,4) \geq 13$. By Lemma 3, $f(13,4k) \geq 13k$. The total number of stops is $52k$. If each floor is served by at least four elevators, then the number of floors is at most $13k$. If some floor is served by at most three elevators, it can be linked to at most $12k - 2$ other floors. Counting this floor, the building can have at most $12k - 1$ floors. It follows that $f(13,4k) = 13k$. □

Concluding Remarks

We summarize our findings so far for $1 \leq m, n \leq 16$ in Table 1.

The ultimate goal of this study is, of course, to completely determine $f(m,n)$, but the task appears difficult. Some intermediate goals, probably achievable but most likely messy, are the determination of $f(4,n)$, $f(5,n)$, and $f(7,3k+1)$. New constructions for perfect buildings will certainly help.

Part V

Take a Shape

Jordan as a Jordan Curve

Robert Bosch

The *Jordan Curve Theorem* states that every simple (i.e., does not cross itself) closed curve in the Euclidean plane divides the plane into two regions—the part that lies outside the curve, and the part that lies inside it. (It is named after the French mathematician Camille Jordan (1838–1922), who is also known for bringing Galois theory into the mainstream and drawing attention to its importance.) The theorem is one of those results that seems completely obvious, yet is extraordinarily difficult to prove. Figure 1, produced using techniques described in [1] and [2], displays a Jordan Curve of Jordan.

Note that it is not immediately apparent that the curve in Figure 1 is a simple closed curve. (Yet it is!) And it is not immediately apparent that the curve divides the plane into "outside" and "inside" regions. (Yet it does!)

Figure 2 helps a great deal towards seeing that property; it displays the curve (black) together with the outside region (white) and the inside region (gray).

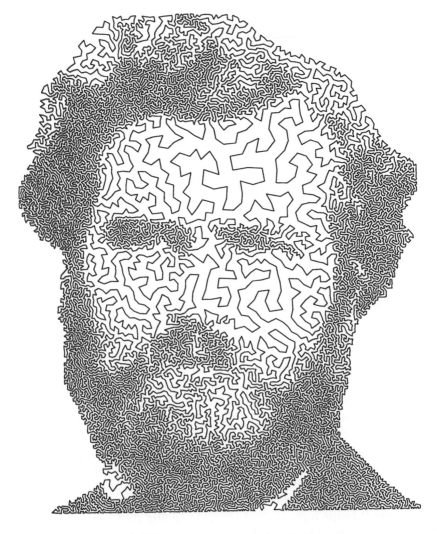

Figure 1. Jordan as a Jordan Curve.

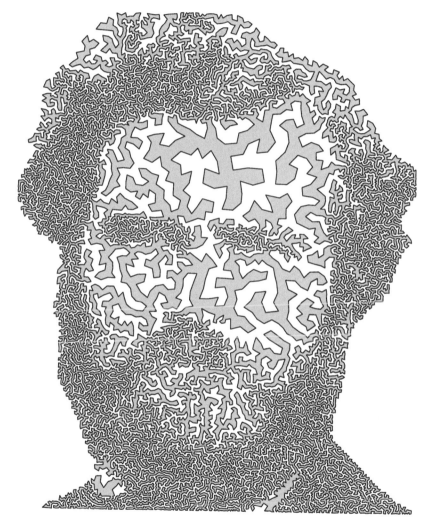

Figure 2. Outside (white) and inside (gray).

Bibliography

[1] R. Bosch and A. Herman. "Continuous Line Drawings via the Traveling Salesman Problem." *Operations Research Letters* 32:4 (2004), 302–
303.

[2] C. S. Kaplan and R. Bosch. "TSP Art." *Proceedings of the 2005 Renaissance Banff Bridges Mathematical Connections in Art, Music, and
Science Conference*, edited by Reza Sarhangi and Robert V. Moody,
pp. 301–308. Winfield, KS: Central Plain Book Manufacturing, 2005.

Wang Tiles, Dynamical Systems, and Beatty Difference Sequences

Stanley Eigen

We seem to be hardwired for repeating patterns. In *The Children of Woot* [6], Jan Vansina studies the Bakuba people of the Democratic Republic of the Congo for whom the invention of a new pattern was considered a major achievement. In fact, every new king was expected to create a new pattern at the outset of his reign and this pattern was displayed throughout the king's reign. There is a classic anecdote about the Bakuba. Missionaries in the 1920s tried to impress the king with a motorcycle. The king had no interest, but when he saw the pattern made by the tire tracks he had it copied and gave it his name.

Today, however, people are interested in patterns that are non-repeating but very, very close to repeating. Here is a new way to tile the plane. I like it because it connects different parts of mathematics and results in a very simple method to create tilings.

Choose any number between $\frac{1}{3}$ and 2. This method then immediately gives an infinite line. For example, if you pick $\frac{1}{2}$ you get the following pattern (which continues to the left and right indefinitely).

With 1 you get this pattern (which also continues left and right).

For $\frac{1}{3}$ you get the next pattern (ditto repeating to the left and right).

These three lines fit together, one on top of another, in a natural way. With a lot more lines, they fill up the entire infinite plane, as shown in Figure 1.

Each of the three lines above are periodic—that is, they have a pattern that repeats over and over. For example, the line for $\frac{1}{2}$ actually consists of the following two squares.

These two are put together into a two-block. And it is this two-block that repeats to the left and right to make the infinite line.

Likewise, the line for $\frac{1}{3}$ is built from these three squares.

These are put together into the three-block.

Figure 1. Portion of an infinite tiling starting $\frac{1}{2}$: Every zigzag row in the infinite tiling is periodic. The top has period two. However, the entire tiling is not periodic. (The period of the bottom row is over $2,000$.)

The line for 1 fits in between the lines for $\frac{1}{2}$ and $\frac{1}{3}$. It is built from three squares forming a three-block.

If you look back to the tiling portion in Figure 1, you will "see" that it consists of "zigzag" rows. Every zigzag row is periodic. That is, it has a repeating block going left and right. Some of the blocks are made of two squares; some are made of three squares. There are also zigzag rows with repeating blocks of lengths 4, 8, 9, 27, and higher powers of 2 and 3 as well as mixed powers, such as 12 and higher. The point is that although every row is periodic, there is no period that works for all rows because the periods get higher and higher.

There is also no repeating pattern going up and down. Again, looking back to the tiling you will see "darker" zigzag rows. These are always separated by either one or two lighter zigzag rows. But the pattern of separation is not periodic.

On the other hand, this is very close to repeating. Every finite pattern in it is repeated over and over again and in all directions. Even more interesting is that it has "pieces" of infinite rows that are not there. For example, the row you get for $\frac{6}{5}$ is

This is built from the five square tiles

that form the five-block that creates the row.

The five-block appears in the full tiling for $\frac{1}{2}$. But the infinite row for $\frac{6}{5}$ does not. This infinite row has period 5, that is, it is built from a repeating five-block. But there is no row in the full tiling for $\frac{1}{2}$ with period 5. The only periods that appear are products of powers of 2 and 3 (like $12 = 2^2 \cdot 3$).

I will conclude by showing you a little of the mathematics behind this—and give you enough information so you can write your own program. For reference, I wrote a Mathematica program in which I need only input a starting number (like $\frac{1}{2}$) and tell it how many rows and colors I want. My source is a pair of papers by the computer scientists J. Kari and K. Culik [2,5].

First let me explain why the row for 1 is directly under the row for $\frac{1}{2}$ and directly above the row for $\frac{1}{3}$. It comes from the following function: if x is between $\frac{1}{3}$ and 1 (including $\frac{1}{3}$), then multiply by 2; if x is between 1 and 2 (including 1) then divide by 3.

$$f(x) = \begin{cases} 2x, & \frac{1}{3} \le x < 1; \\ \frac{1}{3}x, & 1 \le x < 2. \end{cases}$$

As an illustration, $\frac{1}{2}$ is between $\frac{1}{3}$ and 1 so we multiply by 2 and get 1. Hence, the row for 1 is after, or in this case below, the row for $\frac{1}{2}$. This is a dynamical system, and iterating the function forward and backward gives an orbit. Here is a portion of the orbit for $\frac{1}{2}$, which tells you what rows go where. The row for $\frac{3}{2}$ is above the row for $\frac{1}{2}$.

$$\rightarrow \frac{3}{2} \rightarrow \frac{1}{2} \rightarrow 1 \rightarrow \frac{1}{3} \rightarrow \frac{2}{3} \rightarrow$$

By the way, since we are only multiplying and dividing by 2 and 3, there is no way the number $\frac{6}{5}$ or any fraction with 5 on the bottom can appear in the orbit. So the row for $\frac{6}{5}$ will never appear in the tiling for $\frac{1}{2}$.

Next, you should ask where the square tiles come from. These are called *Wang tiles*—squares with colored edges. The rule for putting them together is abutting edges have matching colors. In this example, there are thirteen tiles altogether, as shown in Figure 2.

The side colors (or shades of gray in this case) are irrelevant. Side numbers are produced and then you assign colors to them

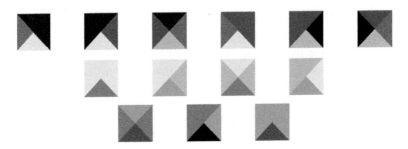

Figure 2. The thirteen tiles.

however you like. The numbers are produced by four formulas— one for each of the four sides. With these formulas and the definition of the function f given earlier, you have enough to write your own program. Here are the four formulas in Mathematica.

- Top formula: `Floor[n*x] - Floor[(n-1)*x]`

- Left formula: `q*Floor[(n-1)*x] - Floor[q*(n-1)*x]`

- Bottom formula: `Floor[q*n*x] - Floor[q*(n-1)*x]`

- Right formula: `q*Floor[n*x] = Floor[q*n*x]`

The number x is just your starting number. The number q is 2 if x is between $\frac{1}{3}$ and 1, and $\frac{1}{3}$ if the number is between 1 and 2. These come from the function f. The number n goes from 1 to how many tiles you want in a row. The `Floor` function truncates the number to the nearest integer below or equal to the number. So, for example, with $x = \frac{1}{3}$ and $n = 11$ you get

$$\texttt{Floor}[\tfrac{11}{3}] - \texttt{Floor}[\tfrac{10}{3}] = 3 - 3 = 0.$$

The top formula gives what is called the *Beatty sequence* for the number x. The bottom formula is the Beatty sequence for the number qx. The side numbers are chosen so that the side colors fit together in a nice row.

If you want to experiment, you can change the function f. I don't know any other formulas for the sides of the tiles. If you find any, let me know.

Bibliography

[1] S. Beatty. "Problem 3173." *American Mathematical Monthly* 33 (1926), 159; solutions in 34 (1927), 159.

[2] K. Culik. "An Aperiodic Set of 13 Wang Tiles." *Discrete Mathematics* 160 (1996), 245–251.

[3] Martin Gardner. "Extraordinary Nonperiodic Tiling that Enriches the Theory of Tiles." Mathematical Games, *Scientific American* 236 (January 1977), 110–112.

[4] B. Grünbaum and G. C. Shephard. Tilings and Patterns, Freeman and Co., N. Y. 1987.

[5] J. Kari. "A Small Aperiodic Set of Wang Tiles." *Discrete Mathematics* 160 (1996), 259–264.

[6] Jan Vansina. *The Children of Woot: A History of the Kuba Peoples.* Madison, WI: University of Wisconsin Press, 1978.

[7] H. Wang. "Proving Theorems by Pattern Recognition." II Bell *System Technical Journal* 40 (1961), 1–41.

The Trilobite and Cross

Chaim Goodman-Strauss

I was a rather smart-mouthed thirteen-year-old and, through my ongoing efforts to distress my elders, had earned for myself many hours of quiet contemplation, which I spent poring over old issues of *Scientific American*. I was lucky enough to know already of "Mathematical Games," and I suppose that was probably why I had the magazines in the first place.

I remember quite distinctly (though perhaps inaccurately) reading about Robert Amman's various aperiodic tiles. This was something astounding: somehow, mysteriously, these tiles could fit together, but only in a way that never quite repeated regularly. Or something like that. I'm quite sure that I really had no idea what I was reading. But I was fascinated, and the memory was sealed.

Only many years later, in graduate school, did I really learn anything about aperiodicity, which only deepened the mysteries, and a large part of my career has been based on studying those subtle little critters.

Aperiodicity is an *astounding* condition, much more subtle than mere non-periodicity. A nonperiodic tiling is simply one that has no periodic symmetry and even mere unadorned rectangles can form nonperiodic tilings. But of course rectangles can also form

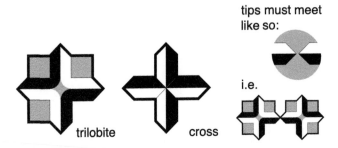

Figure 1. The trilobite (left) and cross (middle) are aperiodic; they can be used to create tilings, but *never* periodically. Where tips of the tiles meet (right), the colors must continue across from one tip to the other.

periodic tilings, and there is nothing especially subtle about the way they can be fitted together.

Aperiodicity is a property of tiles: a set of tiles is aperiodic if they can be used to make tilings, but *only* nonperiodic tilings. That is, somehow, just by fitting together, the tiles can *force* a kind of apparent disorder, allowing no perfect repetition at any scale. From time to time I am still surprised that this is even remotely possible.

Without any further ado, may I introduce to you the trilobite and cross (depicted in Figure 1), one of only a handful of known aperiodic pairs of tiles.[1] They have long existed in a technical paper [4], buried in bound volumes, but I bring them out now to share them with those that might enjoy them the most.

I heartily encourage the reader to engage a copy of Figure 7 or Color Plate IV with a pair of scissors and to experiment with the tiles. One of the salient features of tilings is that it is really pretty difficult to figure out, a priori, what even a simple set of tiles can or cannot do. (In fact, the whole subject of aperiodicity follows the famous result of R. Berger [2], that there is no algorithm, whatsoever, to tell whether a given, arbitrary set of tiles can be used to form a tiling!) Generically, tiling behavior is unfathomable, to the delight of puzzlers everywhere.

---

[1] It is still a famous open question whether there can be a planar tile that by itself is aperiodic. In three-dimensional space, the Schmidt-Conway-Danzer tile can only form tilings with no *translational* symmetry, but it does admit tilings with a periodic screw symmetry, and so might be called *weakly aperiodic* [6].

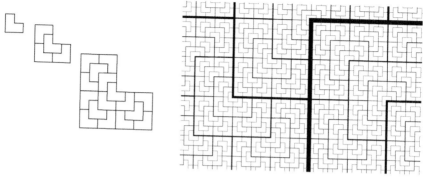

Figure 2. *L*-shaped tiles can be formed into hierarchies of larger and larger *L*-shaped patches; a tiling formed of such hierarchies would be nonperiodic.

But of course, the trilobite and cross were found by coming at the problem the other way round: they were designed with a certain behavior in mind. Like most aperiodic sets known today, they are designed so that they form an infinite hierarchy of larger and larger structures.

Consider an *L-substitution tiling*; one begins with an *L*-shaped tile. Four *L*-shaped tiles can be fitted together to cover an *L*-shaped region exactly twice as large as the original *L*-tile. Four of these *L-supertiles* can be fitted together to make a still larger *L*-shaped region, etc. In this way, we can form *L*-shaped regions of any size; each such region consists of *L*-tiles organized into supertiles, organized into supersupertiles, etc., and in only one way. In fact, we can consider tilings of the entire plane, by *L*-tiles, in which every tile is organized, uniquely, into an infinite hierarchy of larger and larger *L*-shaped patches, as shown in Figure 2.

Even though these *L*-substitution tilings are highly ordered, they are nevertheless nonperiodic. Any supposed period would be have to be at least as wide as an *L*-tile, but also (for exactly the same reason) at least as wide as an *L*-supertile, an *L*-supersupertile, etc. No finite shift could ever be a period for an *L*-substitution tiling.

But *L*-shaped tiles in themselves are not aperiodic; they can form periodic tilings just as easily as nonperiodic ones. The trilobite and cross were designed so that they can *only* be fitted together into larger and larger *L*-shaped patches, to emulate, in a

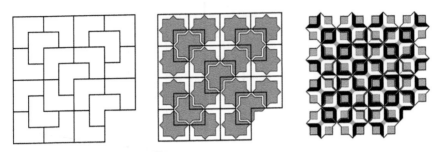

Figure 3. The trilobite and cross can only fit together to form such hierarchies.

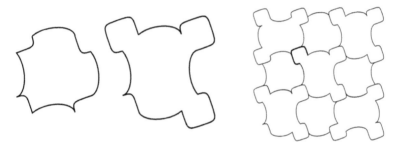

Figure 4. A variation on the trilobite and cross that requires only that the tiles fit together.

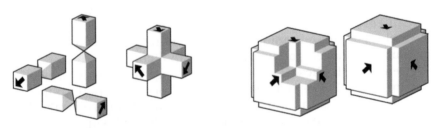

Figure 5. A three-dimensional aperiodic pair of tiles; copies of the "double-obelisk" (left) fit together to make an analogue of the cross. The "armchair" tile (right, shown in two views) is an analogue of the trilobite. The arrow markings are to match when two tiles touch. Similar tiles can be constructed in all higher dimensions.

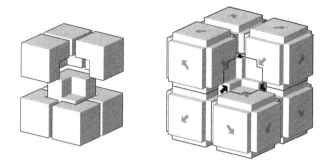

Figure 6. Eight three-dimensional *L*-tiles fit together to make a larger *L*-tile (left) and can form tilings of space by hierarchies of larger and larger *L*-supertiles; the armchair and double-obelisk can only fit together to form larger and larger *L*-shaped regions, and so are aperiodic.

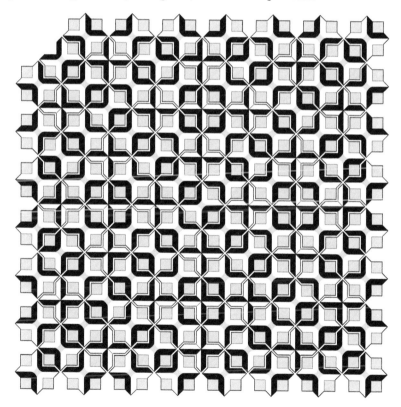

Figure 7. An aperiodic tiling of the trilobite and cross. (See Color Plate IV.)

sense, the L-substitution tilings. (See Figure 3.) To prove this, one shows that trilobites and crosses can only fit together to make a "super-trilobite" and "super-crosses"—patches of tiles that themselves can fit together just as trilobites and crosses must. These in turn can only form supersuper-trilobites and crosses, and so on. The trilobite and cross are aperiodic!

Purists may complain that the tip-to-tip rule is a little nonstandard. A modification of the trilobite and cross, shown in Figure 4, requires only that the tiles fit together. But the proof that the modified tiles behave as advertised is remarkably tedious and involved.

Finally, we note that the trilobite and cross generalize to all higher dimensions! There are a pair of tiles, in any n-dimensional Euclidean space, which emulate a higher dimensional analogue of the L-substitution tiling and so, too, are aperiodic. See Figures 5 and 6.

Bibliography

[1] R. Ammann, B. Grunbaum, and G. C. Shepherd. "Aperiodic Tiles." *Discrete and Computational Geometry* 8 (1992), 1–25.

[2] R. Berger. *The Undecidability of the Domino Problem,* Memoirs of the American Mathematical Society 66. Providence, RI: American Mathematical Society, 1966.

[3] Martin Gardner. "Extraordinary Nonperiodic Tiling that Enriches the Theory of Tilings." *Scientific American* 236 (1977), 110–121.

[4] Chaim Goodman-Strauss. "A Small Aperiodic Set of Planar Tiles." *European Journal of Combinatorics* 20 (1999), 375–384.

[5] Chaim Goodman-Strauss. "An Aperiodic Pair of Tiles in E^n for all $n \geq 3$." *European Journal of Combinatorics* 20 (1999), 385–395.

[6] Chaim Goodman-Strauss. "Open Questions in Tilings." Notes available at http://comp.uark.edu/~strauss/papers, 2000.

[7] R. M. Robinson. "Undecidability and Nonperiodicity of Tilings in the Plane." *Inventiones Mathematicae* 12 (1971), 177–209.

Orderly Tangles Revisited

George W. Hart

In the 1970s and 1980s, Alan Holden described symmetric arrangements of linked polygons that he called *regular polylinks* and constructed many cardboard and stick models. The fundamental geometric idea of symmetrically rotating and translating the faces of a Platonic solid is applicable to both sculpture and puzzles. The insight has been independently discovered or adapted by others, but the concept has not been widely used because no closed-form method is known for calculating the dimensions of snugly fitting parts. This paper describes a software tool for the design and visualization of these forms that allows the dimensions to be determined. The software also outputs geometry description files for solid freeform fabrication and image files for printing paper templates. Paper templates make it easy to teach the concepts in a hands-on manner. Examples and variations are presented in the form of computer images, paper, wood, metal, and solid freeform fabrication models.

Figure 1. Rotation and translation of cube faces.

Figure 2. Regular polylink with six squares.

Introduction

Figures 1 and 2 illustrate the key ideas of Holden's *regular polylinks* [6–8]. The six hollow faces of the inner cube are separated, translated radially outward from the center, and each rotated the same angle clockwise about their center. In Figure 2, the original cube is removed, the faces are moved inward until they interweave, the rotation angle is adjusted slightly, the thickness of each plane is reduced to paper-thinness, and the size of each square hole is shrunk slightly to make a snug fit. Each square links with four others. From the underlying cube's symmetry, the rotational axes are preserved, but not the mirror planes.

It is instructive to create a paper model of Figure 2 by cutting out six hollow squares and interweaving them. Four of the squares can be cut open, linked through the other two, and then taped together. Figure 3 is a photo of a model made using three colors of card stock, iso-coloring parallel faces. It is chiral so one must choose between two enantiomorphs. The only critical parameter is the ratio of the edge length of the outer square to the inner square hole. By means of the software described below, a ratio of approximately 15/11 is determined to be suitable. So the outer square can be 3.7 inches on edge with a 2.7-inch square hole, leaving 0.5 inch of solidity on four sides, and four fit within an 8.5 by 11 inch sheet. This is easily drafted and cut out from paper. Or simply

Figure 3. Paper model of six linked squares.

Figure 4. Hollow square template for making the model in Figure 3.

make six enlarged photocopies of Figure 4. For larger models, use cardboard. Figure 5 has the same weaving as Figure 3, but the strut cross section is made into squares, to give it enough internal substance to hold together when fabricated as a solid freeform fabrication model. This fused deposition model (FDM) is 2.5 inches in diameter.

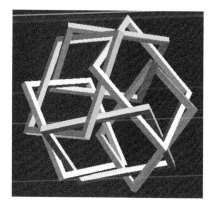

Figure 5. FDM model with square strut cross sections.

Figure 6. A different weaving of six squares, also based on the cube.

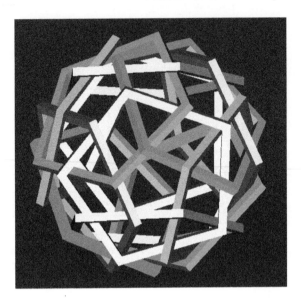

Figure 7. Regular polylink with twelve pentagons.

Holden's regular polylinks are the topologically distinct polygon linkages that result from varying the rotation and radial translation starting with the faces of any of the five Platonic solids. There can be several distinct ways of linking the rotated faces from any underlying polyhedron. Figure 6 shows a second way of linking six squares, but they still lie in the face planes of an imagined cube. Starting from a dodecahedron, one regular polylink of 12 pentagons is shown in Figure 7.

The struts that form the polygon edges in Figures 1–3 have rectangular cross sections, and in Figures 5–7 they have square cross sections. Holden made his models using 1/4-inch diameter wooden dowels of circular cross section. The critical dimension for a snug fit is the ratio of the strut length to diameter. He experimented until he found the shortest lengths that could be assembled, and gave a table of his results for others who replicate the constructions with round dowels [8]. But different dimension ratios are needed if one prefers square, rectangular, or other cross sections.

Figure 8 illustrates a special case to consider. The faces of a dodecahedron are translated into a distance of zero from the cen-

Figure 8. Regular polylink with six pentagons.

Figure 9. Paper model of Figure 8.

ter. Then by proper choice of rotation angle, it is possible to cause opposite faces of the polyhedron to coincide. So, the twelve faces of the dodecahedron fuse into six concentric interwoven pentagons of Figure 8. A paper model of this, Figure 9, is made from pentagons of 3-inch edge, having pentagonal holes of 2.5-inch edge. Analogously, the six faces of a cube can be translated to the origin where they fuse into three orthogonal concentric squares. But because squares have an even number of sides, they intersect other squares (in either of two rotations) rather than forming a weave. With pentagons or triangles a non-self-intersecting polylink can be formed.

The symmetric intricacy attainable with simple components makes polylinks very appealing aesthetically. Holden illustrates only small cardboard models and dowel models, but he suggests their use in "constructivist sculpture." I have run into a dozen or so examples of sculptures based on his suggestion or a rediscovery of the essential ideas, and there are many ways to adapt, combine, or extend them. For example, Bill Chertoff, Robert Lang, George Odom, Rinus Roelofs, and Carlo Sequin have explored the minimalist construction of four triangles shown in Figure 10 [1, 2, 10–13]. It can be derived from either the tetrahedron or the octahedron by translating the triangles into the origin. Deriving Figure 10 from a tetrahedron shows that a degree of rotational freedom re-

Figure 10. Regular polylink composed of four triangles.

Figure 11. Coxeter, 1992, with four-triangle model by Odom. (See Color Plate II.)

mains. Starting from the octahedron explains its axis of fourfold symmetry; a particular rotation angle causes faces to merge in pairs. H.S.M. Coxeter analyzed this construction and showed: (1) if made of zero-thickness material, the hole in each triangle has exactly half the edge length of the whole triangle, and (2) the 12 outer vertices lie at the midpoints of a cube's edges, i.e., the vertices of an Archimedean cuboctahedron [3]. Figure 11 shows Coxeter holding the cardboard model that sparked his investigations, sent to him by George Odom.

A very large polylink sculpture is Charles Perry's 1976, 12-ton *Da Vinci*, based on the six-pentagon polylink of Figure 9. (See [4, Plate D] for a figure.) Perry made flat steel pentagons 20-feet on a side, and nested two copies of the construction together. Figure 12 illustrates this idea but takes it further to have three concentric copies. The series can be extended inward to any depth because the components are progressively reduced in size geometrically toward the center.

Holden does not define *orderly tangle* precisely, but uses it loosely to subsume a variety of interesting forms such as highway interchanges, woven cloth, and polylinks. A form such as Figure 12 is not a regular polylink (because it is three regular polylinks) but it fits under the broader heading of orderly tangle.

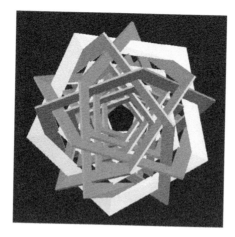

Figure 12. Three nested regular polylinks, each like the one in Figure 8.

Puzzles

A wooden puzzle based on the twelve pentagons of Figure 7 is shown in Figure 13. In Figure 14 is a wooden puzzle made of 30 identical sticks, which form ten triangles. It is based on an icosahedron, with the 20 faces translated to the center and rotated to coincide into ten pairs. In both puzzles, the square wooden sticks are cut longer than the polygon edges, and are notched to lock together gluelessly with half-lap joints. I received these beautiful works as presents during a trip to Taiwan in July 2004. They were cleverly designed by Teacher Lin and expertly constructed by Sculptor Wu, both members of the Kaohsiung Puzzle Club. I received them disassembled for easy transport in my luggage and had to assemble them on arriving home.

This puzzle idea of Lin and Wu can be applied to many other polylinks if one can determine the proper length to cut the wooden sticks. As a simple example, the polylink shown in Figure 5 can be built from square stock if the ratio of the outer edge to the cross-section edge is 9.8 to 1. This value comes from the software described in the next section, but as wood is flexible and compressible, some experimenting was still required. So from 1-unit square stock, one can cut 24 pieces, each 11.8 units long. Near each end, but from opposite sides, notch halfway through to leave a 1-unit overhang beyond the notch. Figure 15 shows the assembled

Figure 13. Wooden puzzle by Lin and Wu.

Figure 14. Another wooden puzzle by Lin and Wu.

Figure 15. Puzzle based on Figures 3 and 5.

Figure 16. Puzzle based on Figure 8.

result, built of 0.5-inch square wood bars. Figure 16 is the analogous puzzle of thirty sticks assembled into six pentagons woven as in Figure 8, built of 3/8-inch aluminum bars.

With wood, I found that cutting the lengths a few percent shorter than the software suggests seems to compensate for the flexibility of the struts and the fact that the corners of wood struts may be compressed. So although the software provides a good initial value, the woodworker is advised to plan on some experimentation with scrap wood before investing in quality wood.

Software

To design and build regular polylinks and tangles of concentric polylinks, I wrote a program with sliders that can be adjusted to see a wide range of structures onscreen. The computer-generated images in this paper are from screenshots of its operation. The user can specify any number of polylinks to be assembled concentrically, and for each set, the user selects the underlying polyhedron used as its basis. Then sliders allow the size, translation, and rotation of the components to be adjusted. At present, the edge cross sections are adjustable rectangles, allowing flat (paper) versions as the special case of width zero. As the sliders are adjusted, the dimensions of length, width, thickness, etc. are displayed, which can be used for making models of wood or other materials. When the user is happy with the form shown rotating on the screen, a click of a button generates an STL file for making solid freeform fabrication models, e.g., Figure 5. Clicking another button generates an image file that can be printed for making paper or cardboard models, e.g., Figure 4.

Three 4-cm examples made by selective laser sintering (SLS) from the program's STL output are shown below. Figure 17 is ten triangles, arranged as in Figure 14, but with a "tall" cross section. Figure 18 is 20 icosahedrally arranged triangles, with the minimum possible linkage (analogous to the cubic form in Figure 6). Figure 19 is a weave based on the octahedron but with the triangles replaced by hexagons.

The software is freely available online at my website [5]. It has only been tested on PCs, but it is written in java so should be transportable to other computing environments. For rendering three-dimensional images on the screen, it uses Sun's freely available *java3D* extension, which must be installed on the user's computer. The coding is quite straightforward, with no clever algorithms or data structures needed. I believe it can be straightforwardly replicated by any software engineer with expertise in graphics programming.

Users may verify their understanding of its capabilities by replicating the figures of this paper, or one can simply start playing to create new objects. The software generates all regular polylinks

Figure 17. Ten triangle SLS model, based on icosahedron.

Figure 18.   Twenty triangle SLS model, based on icosahedron.

Figure 19. Eight hexagon SLS model, based on octahedron.

including several icosahedral examples that Holden did not describe. Lang enumerated the regular polylinks with a computer search and showed there are two tetrahedral, two cubic, three octahedral, five dodecahedral, and 23 icosahedral varieties (excluding the extreme cases where polygons either are not linked or are merged into pairs) [9]. I have replicated these numbers with an independent geometric technique of counting subsets of symmetry axes that pierce the interior of a polygon centrally placed in the stellation diagram.

Many extensions of the program are possible. It could be modified in a straightforward way to produce circular, triangular, or other shapes of cross sections for the polygon edges. Another possible addition to the software is a numerical search for slider settings that result in a snug fit. I planned for this feature when originally designing the software, but then discovered that it is simple to see on the screen if there are gaps or overlaps in the components, so manual adjustment seems sufficient.

Variations

There are an unlimited number of variations on the above ideas. For example, Holden made several models of linked polygonal stars instead of convex polygons. Similarly, one might try rectangles or rhombs instead of regular polygons; I have not implemented any of

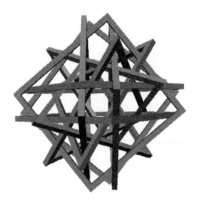   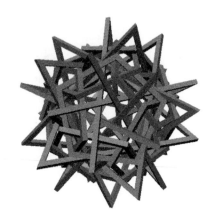

Figure 20. Six squares plus four triangles.   Figure 21. Twelve pentagons plus ten triangles.

these with software. (For related rectangle linkages, see [12].) Another idea is to start with polyhedra other than the Platonic solids. Holden constructed examples based on Archimedean solids, e.g., the rhombicuboctahedron and the snub cube.

One can symmetrically combine multiple polylinks based on different polyhedra. Sets with the same symmetry combine to form a tangle with the same symmetry. Figure 20 shows the combination of a cubical and a tetrahedral form. It combines the tangles of Figure 2 and Figure 10 so each triangle links three squares while each square links two triangles. Figure 21 shows the union of a dodecahedral and an icosahedral form in which both puzzles of Figures 13 and 14 are intertwined.

Another variation Holden tried was to put squares around the twelve twofold axes of a cube, i.e., in the planes of a rhombic dodecahedron. Here the squares only exhibit twofold symmetry, so their edges and vertices are not all equivalent, and the polylink is not regular. I find the results less attractive—fun to generate with the software but less interesting to look at. An example is shown in Figure 22. Here the 12 squares are grouped together as the sides of four triangular prisms. Extending his idea, it is natural to try putting 30 squares around the twofold axes of an icosahedron, i.e., in the planes of a rhombic triacontahedron. These can be grouped in many interesting ways, including five cubes or six pentagonal

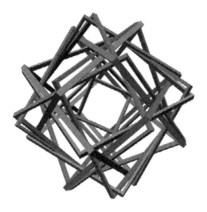

Figure 22. Twelve squares in four triangular prisms.

Figure 23. Thirty squares in ten triangular prisms.

prisms. In Figure 23, the 30 squares are grouped together as the sides of ten triangular prisms. (When brought to the origin and fused into 15 squares, they form five octahedra.)

Spiraling the faces in a series of concentric arrangements is another idea that can be explored with the software. Figure 24 shows a simple example based on an inner cube that is surrounded by nine layers, each slightly larger and rotated five degrees more than the previous. At the end of nine steps, the 45-degree rotation of the squares leads them to be arranged as in a cuboctahedron. This is not technically a "tangle" as no polygons are linked, but it is easy to generate other examples that are. Figure 25 shows an assemblage of triangles that smoothly rotate between an outer stella octangula and an inner form of four triangles, positioned as in Figure 4. Figure 26 shows a linked set of ten triangular helixes that each join a pair of opposite faces of the outer icosahedron. Charles Perry has explored this "face screwing" concept to great effect in a number of monumental sculptures, e.g., his 1973, 35-foot tall, *Eclipse*, which starts with an inner dodecahedron. (See [4, Plate D] for a figure.)

Weaving is another way to generate variations on a polylink. The oldest example that I know of a form closely related to a regular polylink is the sepaktakraw ball, seen in Figure 27. Traditionally woven from rattan, it exhibits the dodecahedral pattern

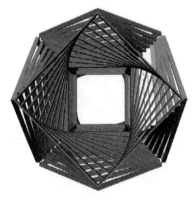

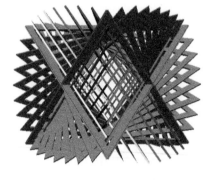

Figure 24. Ten sets of squares: inner cube, outer cuboctahedron.

Figure 25. Set of triangles join opposite faces of stella octangula.

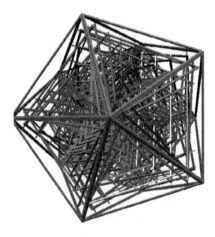

Figure 26. Sets of triangles join opposite faces of icosahedron.

of Figure 8. The design goes back centuries for use in traditional
Asian "football" games. Figure 28 shows a "spherical basket" I wove
of paper strips. The six dark central bands follow the same weave
pattern of six pentagons shown in Figure 8. Neighboring bands,
which get progressively lighter in color, simply alternate over and
under in the natural weave pattern. Note the difference between
this weave and the sepaktakraw ball. In Figure 28 the individual

Figure 27. Sepaktakraw ball.

Figure 28. Paper weave. (See Color Plate III.)

Figure 29. Skew holes.

strands weave, while in Figure 27 the group of strands weaves as a whole.

Another variation, illustrated in Figure 29, is to allow a parameter for a relative rotation between the polygon and its hole. This adds a dynamic visual quality to the forms.

A different type of variation is to replace the straight edges of the polygons with curved paths. There are infinitely many ways to choose curves, but a particularly natural one is to perform an

Figure 30. Inversion of four triangles.

inversion about the center of symmetry. Central inversion replaces each point at distance r from the origin with a point in the same direction but at distance $1/r$. This transformation is well studied mathematically but little used in sculpture [13]. In this context, the chain of n rectangular struts that form the edges of an n-gon is replaced by a chain of n curved volumes bounded by four portions of spheres. Figure 30 shows an example in which a square-cross-section version of the four-triangle construction in Figure 10 is inverted into a structure that resembles four interlocked three-leaf clovers. The 12 corners are nowon the inside, but remain

Figure 31. Five tetrahedra, an icosahedral polylink.

Figure 32. Inversion of five tetrahedra.

60-degree angles because central inversion is an angle-preserving transformation.

Figure 31 is the well-known compound of five regular tetrahedra, which is easily generated as an icosahedral polylink. It is formed here with struts of rectangular cross section that overlap to make ribbed edges. These invert into the labial forms of Figure 32. The interior regions of Figures 30 and 32 are very interesting spaces, difficult to capture in a still image.

Conclusion

Regular polylinks are a rich source of fundamental forms that may be used as the basis for a gamut of three-dimensional design ideas. Alan Holden's 1983 book abounds with creative inspirations displaying their symmetric elegance. But the spectrum of examples presented there and expanded upon here only scratch the surface. To introduce polylinks in a concrete manner, paper constructions such as Figures 3, 9, and 11 make a good hands-on educational activity. After that, I hope that the polylink-generation software described here will enable readers to explore new possibilities.

Acknowledgments. Portions of this paper appeared at the Bridges 2005 conference, "Renaissance Banff." Thank you *RJT Educational Training* for making the model of Figure 5, and thank you Jim Quinn for making the models of Figures 17–19. Figure 11 is by Marion Walter.

Bibliography

[1] H. Burgiel, D. S. Franzblau, and K. R. Gutschera. "The Mystery of the Linked Triangles." *Mathematics Magazine* 69 (1996), 94–102.

[2] W. W. Chernoff. "Interwoven Polygonal Frames." *Discrete Mathematics* 167–168 (1997), 197–204.

[3] H. S. M. Coxeter. "Symmetric Combinations of Three or Four Hollow Triangles." *Mathematical Intelligencer* 16 (1994), 25–30.

[4] Michele Emmer, editor. *The Visual Mind.* Cambridge, MA: MIT Press, 1993.

[5] George W. Hart. http://www.georgehart.com, 2008.

[6] Alan Holden. *Shapes, Spaces and Symmetry*. New York: Columbia University Press, 1971. (Dover reprint, New York, 1991.)

[7] Alan Holden. "Regular Polylinks." *Structural Topology* 4 (1980) 41–45.

[8] Alan Holden. *Orderly Tangles: Cloverleafs, Gordian Knots, and Regular Polylinks*. New York: Columbia University Press, 1983.

[9] Robert J. Lang. "Polypolyhedra in Origami." In *Origami³*, edited by Thomas Hull, pp. 153–167. Natick, MA: A K Peters, 2002.

[10] Rinus Roelofs. http://www.rinusroelofs.nl, 2008.

[11] Doris Schattschneider. "Coxeter and the Artists: Two-Way Inspiration." In *The Coxeter Legacy: Reflections and Projections*, edited by C. Davis and E. W. Ellers), pp. 255–280. Providence, RI: American Mathematical Society, 2005.

[12] Carlo Sequin. "Analogies from 2D to 3D: Exercises in Disciplined Creativity." In *Proceedings of the 1999 Bridges Conference on Mathematical Connections in Art, Music, and Science, edited by Reza Sarhangi*, pp. 161–172. Winfield, KS: Central Plain Book Manufactoring, 1999. Also *Visual Mathematics* 3:1 (2001), http://www.mi.sanu. ac.yu/vismath/sequin1/index.html.

[13] John Sharp. "Two Perspectives on Inversion." In *Meeting Alhambra: Proceedings of ISAMA-Bridges 2003*, edited by Javier Barrallo et al., pp. 197–204. Granada, Spain: University of Granada, 2003.

Quasi-Periodic Essays in Architectural and Musical Form

Akio Hizume

Since 1983 I have been researching quasi-periodic patterns based on the Fibonacci Lattice, the Penrose Lattice, and the Golden Mean. I have made many quasi-periodic structures, all named "Star Cage," for one-, two-, and three-dimensional spaces. These designs can be applied both to the architecture of the future and to music. We can actually live in such mathematical structures with delight. Shown here are a few of my works.

In 1990, I published drawings of a quasi-periodic architecture based on the Penrose Lattice (Figure 1) [4]. When I designed it, I happened to find a five-fold weaving pattern (Figure 2). In 1992, I expanded this quasi-periodic weaving pattern into a three-dimensional structure that is related to one of Stewart T. Coffin's puzzles (Figure 3) [5]. I have built many models out of bamboo as public sculptures. Figures 4–8 show some of the variety of small stars, mostly consisting of 30 units each, that I have made.

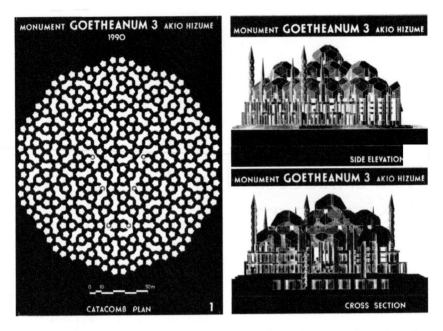

Figure 1. *Goetheanum 3* (1990): quasi-periodic architecture based on the Penrose Lattice.

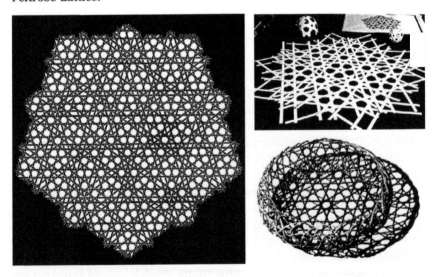

Figure 2. *Go-Magari* (1990): quasi-periodic fivefold weaving based on the Penrose Lattice.

Figure 3. *Mu-Magari* (1992–2003): quasi-periodic three-dimensional chiral lattice based on the Penrose Lattice. (Patent No. 3493499.)

Figure 4. *Pleiades* (1995) consists of six woven pentagonal stars. (Patent No. 3114086.)

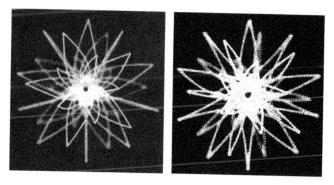

Figure 5. Some examples of *Pentagonal Gravity* (1997): each one consists of 30 coved lines.

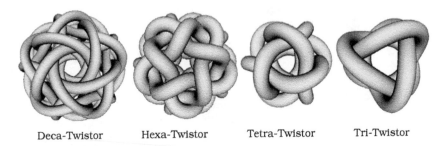

Deca-Twistor   Hexa-Twistor   Tetra-Twistor   Tri-Twistor

Figure 6. *Poly-Twistor Simplest* (2002).

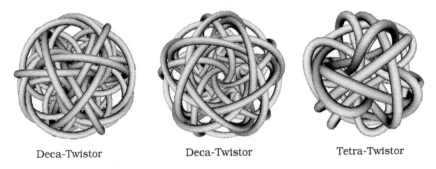

Deca-Twistor   Deca-Twistor   Tetra-Twistor

Figure 7. Some examples of diverse *Poly-Twistor.*

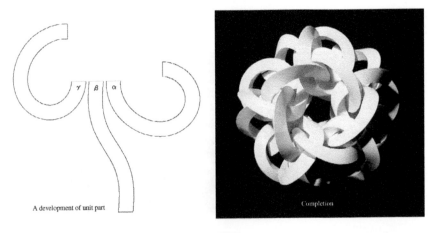

A development of unit part   Completion

Figure 8. *Hexa-Twistor Triangular Section* (2003) is made from paper.

After I learned about phyllotaxis in 1985 from a paper by Kazuo Azukawa [1], I designed a tower structure called *Sunflower Tower* [4], which was actually constructed in 2003 (Figures 9 and 10). This design can also be used as a tunnel structure (Figure 11). I call it *Fibonacci Tunnel.*

In 2005, I drew several kinds of spiral webs based on phyllotaxis (Figure 12). They all consist of similar triangles only. There are single spiral, double spirals, triple spirals, five spirals, eight spirals, thirteen spirals, and so on. Later, I made the same structure out of origami (Figure 13), borrowing Tomoko Fuse's technique called "Nejire Taju Toh" (2002), which means "twisted multiple tower" [3]. At the same time I also made a Fibonacci Tunnel by means of origami (Figure 11, right).

The Fibonacci Lattice is the shadow of a six-dimensional regular lattice projected onto one-dimensional space. When I first learned about it in 1986, I immediately played the quasi-periodic rhythm as music. I designed a staircase using the rhythm of the Fibonacci Lattice at the Goetheanum 3 (Figure1). I called the steps *Democracy Steps* because they are completely fair for both right and left legs. I actually built these staircases in Japan, U.S.A., and New Zealand as public sculptures (Figure 14).

The Golden Mean is the most dissonant ratio among all possible ratios of two real numbers. It could become a new standard of harmony. I developed a new tone and scale system. In 1992, I discovered how to extend such quasi-periodic rhythm to any irrational number, and then I composed and played a variety of quasi-periodic music. In 1999, I developed an interactive computer rhythm and tone-scale generator based on an arbitrary real number (Figures 15 and 16).[1] Figure 16 explains the scale and spectrum generated by an example based on the square root of 3. In this system, a real number is expressed as a continued fraction and sounds a distinctive tone and tonal scale according to the number selected. With scale and tone integrated perfectly, the Pythagorean Comma, which has been a tough issue for musical theorists for more than 3,000 years, is banished, and you develop an entirely new understanding of harmony.

My website includes detailed descriptions and histories of the evolution of many of my productions: http://starcage.org.

---

[1] I presented 275 copies of a CD-ROM containing the interactive rhythm generator, called "Real Kecak System," as an exchange gift at G4G7, Atlanta, GA, 2006.

Figure 9. *Sunflower Towers mod F* (*F* = Fibonacci Number) (1986–2004): they are based on phyllotaxis. (Regarding the relation between the phyllotaxis and congruent expression, refer to [5].)

Figure 10. *Bamboo Sunflower Tower Catenary mod 2* (2004).

Figure 11. *Bamboo Fibonacci Tunnel mod 2* (2004) and origami Fibonacci Tunnels (2005).

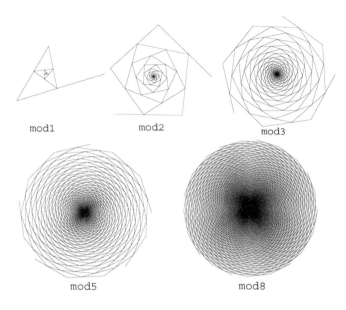

mod1 mod2 mod3

mod5 mod8

Figure 12. A Fibonacci Tornado consists only of similar triangles (2005).

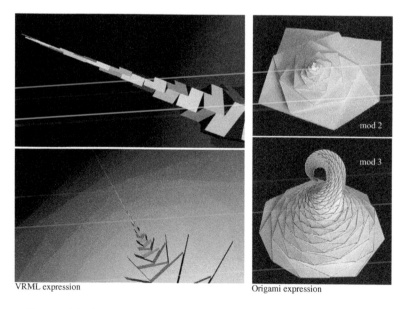

VRML expression Origami expression

Figure 13. Some expressions of the Fibonacci Tornado (2005).

Figure 14. *Democracy Steps* (1990–2005): quasi-periodic staircase based on the Fibonacci Lattice.

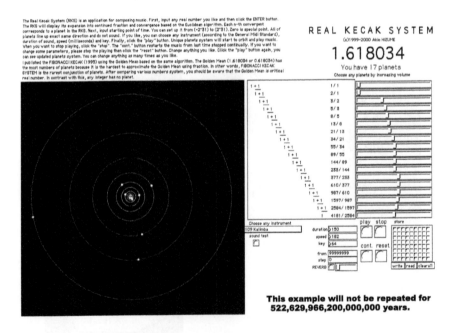

Figure 15. *Real Number Music* (1999–2000): interactive quasi-periodic poly-rhythm generator.

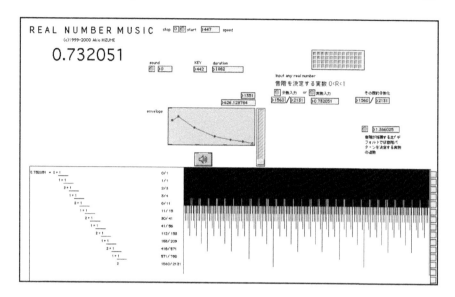

Figure 16. *Real Number Music* (1992–2002): interactive quasi-periodic tone and scale generator.

Bibliography

[1] Kazuo Azukawa. *Sunflower Seeds*, Sugaku Seminar 7. Tokyo: Nihon Hyoron-sha, 1985.

[2] Stewart T. Coffin. *The Puzzling World of Polyhedral Dissections*. Oxford, UK: Oxford University Press, 1990. Second edition: *Geometric Puzzle Design*. Wellesley, MA: A K Peters, 2006.

[3] Tomoko Fuse. "Twisted Multiple Tower." *Manifold* 5 (2002), 3–5.

[4] Akio Hizume. *Life and Architecture*. Gunma: Akio Hizume, 1990.

[5] Akio Hizume. "Aperiodic Space Structure with Straight Member Assembled in Sextuple Form." Patent no. 3493499. March 31, 1993.

[6] Koji Miyazaki. *Polyhedron and Architecture*. Tokyo: Sho-Koku-sha, 1979.

[7] Tohru Ogawa. *3D Penrose Transformation*, Sugaku Seminar 1 and 2. Tokyo: Nihon Hyoron-sha, 1986.

[8] Roger Penrose. "Pentaplexity." *Mathematical Intelligencer* 2 (1978), 32–37.

Ellipses

Robert Barrington Leigh, Ed Leonard,
Ted Lewis, Andy Liu, and George Tokarsky

A vertical wall and a horizontal floor meet at the point X. A vertical ladder is standing against the wall. A cat jumps onto it and sits at its midpoint C. This causes the ladder to move, with its apex A sliding along the wall down toward the floor, and its bottom B sliding along the floor away from the wall. The whole motion takes place in a vertical plane and comes to a stop when the ladder is lying horizontally on the floor. (See Figure 1.) What is the curve traced by the cat?

Such problems are called *loci* problems. The set of all possible positions of a point satisfying certain conditions is called the *locus of the point*. For instance, the locus of a point at a fixed distance from a fixed point is a circle, while the locus of a point at a fixed distance from a fixed line is a pair of lines, one on each side of the given line, and parallel to it.

In the problem with the cat on the sliding ladder, complete the rectangle $AXBY$. Then C is also the midpoint of the other diagonal XY. Although A and B are variable points, the length AB is fixed, and so is the length XY. Since CX is half this length, C is at a fixed

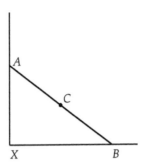

Figure 1. The ladder, AB, with the cat at midpoint C.

distance from the fixed point X. Hence, its locus is a quarter of a circle.

Suppose the point C where the cat sits is not the midpoint of AB, but satisfies $AC = 2BC$. If the ladder moves in the same way as before, what is the curve traced by the cat then? To answer this question, we need to know more loci. One of the best books on this subject is the one by Gutenmacher and Vasilyev [2, pp. 1–8], from which the above problem comes. However, we obtain the answer in a different way.

Let F and D be fixed points. What is the locus of a point P such that $PF/PD = e$, where e is an arbitrary positive constant? For the special case $e = 1$, the locus is the perpendicular bisector of FD. If $e \neq 1$, the locus is known as an *Apollonius circle* with respect to F, D, and e. We may assume that $e < 1$. If $e > 1$, just interchange the labels for the points F and D.

We present a construction of this Apollonius circle. (See Figure 2.) Construct triangle PDF where $PD = DF$ and $PF = eDF$. Bisect the interior and exterior angles at P, cutting the line DF at the points V and V', respectively. The circle with diameter VV' is the desired Apollonius circle. We leave the justification as an exercise.

The distance between a point P and a line d, denoted by Pd, is the length of the segment PD where D is the point on d such that PD is perpendicular to d. If P is on d, then it coincides with D and $Pd = 0$.

Let f and d be fixed lines. What is the locus of a point P such that $Pf/Pd = e$, where e is an arbitrary positive constant? In the case where f and d are parallel, it is a line between them and

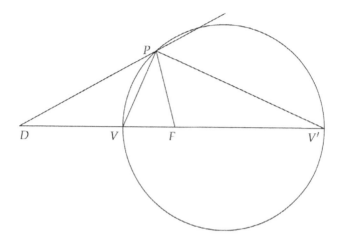

Figure 2. Apollonius circle.

parallel to them. In the case where f and d intersect, the locus consists of two lines through their point of intersection.

Let F be a fixed point and d a fixed line. What is the locus of a point P such that $PF/Pd = e$, where e is an arbitrary positive constant? For the special case $e = 1$, it is a curve not encountered in elementary Euclidean geometry. It is called a *parabola*, a member of a type of curves belonging to a family known as the *conic sections*. These curves are usually studied in projective geometry or analytic geometry, but see [3] for a Euclidean look at the parabola.

Suppose $e \neq 1$. Because of the nonsymmetry between F and d, the nature of the locus depends on whether $e < 1$ or $e > 1$. The results are the other two types of conic sections. If $e < 1$, the curve is called an *ellipse*. If $e > 1$, the curve is called a *hyperbola*. The point F is called a *focus*, the line d a *directrix*, and the constant e the *eccentricity* of the curve.

Although the ellipses and the hyperbolas look very different in the Euclidean plane, they do have many common properties. In the projective plane, the distinction between them vanishes, but that is another story. In this paper, we will take a look at the ellipse from within Euclidean geometry. It is not difficult to modify our approach and apply it to the hyperbola. We leave that as an exercise.

Construction of Points on an Ellipse

Let us state from the outset that an ellipse is not a straight line and not a circle, even though a circle may be considered as a limiting case of an ellipse. That an ellipse is not a straight line will become evident very soon. We will show eventually that it is not a circle, either.

The Euclidean tools of straightedge and compass allow only the construction of lines and circles. Thus, we cannot construct an ellipse within Euclidean geometry. What we will do is construct the points of intersection of an ellipse with an arbitrary straight line. Many of the important properties of the ellipse will emerge from this exercise. However, see [1, pp. 173–183] for some non-Euclidean methods for constructing an ellipse, plus its other interesting properties not considered here.

We are given the focus F and the directrix d of an ellipse. We take d to be horizontal and F above it. The eccentricity e is given by a segment of length e, shorter than another given segment that is taken to be of unit length. In each of the constructions below, the points of intersection of the ellipse with a specific line must also be the points of intersection of that line with some circle. Since a line and a circle intersect in zero, one, or two points, we have indeed found all of them. We will not repeat this for each construction.

We first consider the points of intersection of our ellipse with vertical lines. Denote by a the one through F, intersecting d at the point D. Construct the Apollonius circle with respect to D, F, and e, intersecting a at V and V', with V closer to d. Since

$$\frac{VF}{Vd} = \frac{VF}{VD} = e$$

and

$$\frac{V'F}{V'd} = \frac{V'F}{V'D} = e,$$

these points lie on the ellipse.

Now let ℓ be any vertical line ℓ other than a, intersecting d at the point L. Construct the Apollonius circle with respect to F, L, and e. It intersects ℓ in zero, one, or two points. Suppose P is such a point. Then

$$\frac{PF}{Pd} = \frac{PF}{PL} = e,$$

and P lies on our ellipse.

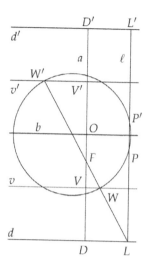

Figure 3. Apollonius circle intersecting FL at W and W'.

Let this Apollonius circle intersect FL at W and W', with W closer to L. (See Figure 3.) Since $WF/WL = e$, triangles FVW and FDL are similar, so that W lies on the horizontal line v through V. Similarly, W' lies on the horizontal line v' through V'. Since the center of this circle is the midpoint of WW', it lies on the horizontal line b through the midpoint O of VV'. It follows that the ellipse is symmetric about b.

We next consider the points of intersection of our ellipse with any horizontal line m. We draw a circle with center F and radius equal to e times the distance between d and m. It intersects m in zero, one, or two points. Suppose P is such a point. Then $PF/Pd = e$ and P lies on our ellipse. From this, it is immediately clear that our ellipse is symmetric about a. Thus we have established that the ellipse is a centrally symmetric figure. We call a its *major axis*, b its *minor axis*, and O its *center*.

Note that V lies on a. Hence, it is the only point on v that belongs to the locus. Similarly, V' is the only point on v' that belongs to the locus. Thus, we expect that the entire ellipse lies between these two lines. Indeed, for any point P on any horizontal line below v, we have

$$\frac{PF}{Pd} > \frac{VF}{Vd} = e.$$

Now let P be a point on any horizontal line above v'. Let PF intersect v' at Q. Then

$$\frac{PF}{Pd} = \frac{QF}{Qd} > \frac{V'F}{V'd} = e.$$

Thus, the vertical dimension of the ellipse is finite. It follows immediately that so is its horizontal dimension.

Finally, we consider the points of intersection of our ellipse with any line that is neither vertical nor horizontal. Let k be such a line passing through F, intersecting d at a point K other than D. Construct the Apollonius circle with respect to F, K, and e and let it intersect ℓ at P and P'. These are the points of intersection of ℓ with the ellipse.

Suppose k does not pass through F. Let it intersect d at a point K. Let Y be the point on k at a distance 1 above d. Draw a circle with center Y and radius e. It may intersect KF in zero, one, or two points. Suppose X is such a point. Let the line through F parallel to XY intersect k at P. Then

$$\frac{PF}{Pd} = \frac{YX}{Yd} = e,$$

and P lies on the ellipse.

We now establish two important properties of tangents to an ellipse. In the preceding construction, suppose X is the only point of intersection of LF with the circle centered at Y. (See Figure 4.) Then KF is a tangent to this circle, and is perpendicular to the radius XY. Now k is a tangent to the ellipse at P. Since PF is parallel to XY, $\angle PFK = 90°$. In other words, *the part of every tangent between the point on the ellipse and the point on the directrix subtends a right angle at the focus.*

Suppose k is tangent to the ellipse at P and intersects d and d' respectively at K and K'. Let the vertical line through P intersect d and d' respectively at L and L'. (See Figure 5.) Then triangles PLK and $PL'K'$ are similar, so that

$$\frac{PK}{PK'} = \frac{PL}{PL'} = \frac{PF}{PF'}.$$

Since $\angle PFK = 90° = \angle PF'K'$, triangles PFK and $PF'K'$ are also similar. This follows from the rarely used *right angle-side-hypotenuse* case for similarity. Hence $\angle FPK = \angle F'PK'$. In other words, *every tangent makes equal angles with the lines joining the point of tangency to the two foci.*

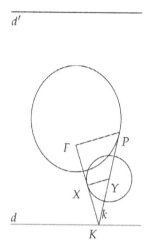

Figure 4. Apollonius circle from Figure 3 with circle centered at Y that intersects FL at X.

An Alternative Definition of the Ellipse

Let F' and D' be the points symmetric to F and D, respectively, with respect to C, and let d' be the horizontal line through D'. Then the same ellipse is defined if we use F' as the focus and d' as the directrix.

Let P be an arbitrary point on the ellipse. Let the vertical line through P intersect d and d' at L and L', respectively. Then

$$\frac{PF}{PL} = e = \frac{PF'}{PL'}.$$

Hence $PF + PF' = e(PL + PL') = eLL'$, which is a constant. This is an alternative definition of the ellipse, as the locus of a point whose distances from two fixed foci have constant sum. With the ellipse defined this way, we can still prove that *every tangent makes equal angles with the lines joining the point of tangency to the two foci.*

Let F and F' be the foci and KK' be a tangent to the ellipse. Let F'' be the reflection of F across KK'. We claim that the point P of intersection of KK' with $F'F''$ is the point of tangency. Since exactly one point on KK' is on the ellipse while every other point on KK' is outside the ellipse, we only need to prove that $PF + PF' < QF + QF'$

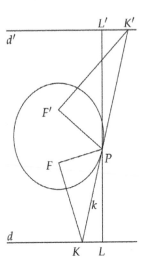

Figure 5. Tangent line making similar triangles.

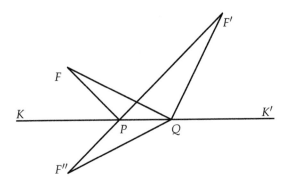

Figure 6. Proof that P is a point of tangency.

for any other point Q on KK'. By the triangle inequality, we have $QF + QF' = QF'' + QF' > F''F' = PF'' + PF' = PF + PF'$ as desired. (See Figure 6.)

The new definition of an ellipse does not mention the directrices at all. Thus, it is impossible for us to prove from it that *the part of every tangent between the point on the ellipse and the point on the directrix subtends a right angle at the focus.* Instead, we shall

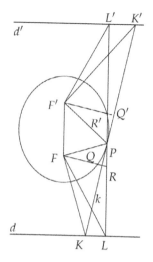

Figure 7. Figure 5 with the addition of Q, Q', R, and R'.

make use of this to define the directrices, and show that the old definition also follows from the new one.

Let P be any point on the ellipse and let k be the tangent to the ellipse at P. Let K and K' be points on it such that $\angle PFK = 90° = \angle PF'K'$. Let d and d' be the horizontal lines through K and K', cutting the vertical line FF' at L and L', respectively. Let the lines through F and F' perpendicular to k cut KK' at Q and Q', and LL' at R and R', respectively. (See Figure 7.)

We claim that d and d' are the desired directrices. For justification, we need to prove first that their choices do not depend on the specific point P on the ellipse. Then we have to show that PF/Pd is constant for all pints P on the ellipse.

Since $\angle FPK = \angle F'PK'$ and $\angle PFK = \angle PF'K' = \angle PQF = \angle PQ'F' = 90°$, triangles PFK, $PF'K'$, PQF, and $PQ'F'$ are all similar to one another. Hence

$$\frac{PK}{PF} = \frac{PK'}{PF'} = \frac{PF}{PQ} = \frac{PF'}{PQ'}.$$

Denote the value of this common ratio by μ. Then

$$\frac{KK'}{PF + PF'} = \frac{PK + PK'}{PF + PF'} = \mu$$

and

$$\frac{QQ'}{PF + PF'} = \frac{PQ + PQ'}{PF + PF'} = \frac{1}{\mu}.$$

Multiplication yields $KK' \cdot QQ' = (PF + PF')^2$.

Since $\angle LPK = \angle L'PK'$ and $\angle PLK = \angle PL'K' = \angle PQR = \angle PQ'R' = 90°$, triangles PLK, $PL'K'$, PQR, and $PQ'R'$ are all similar to one another. Hence

$$\frac{PK}{PL} = \frac{PK'}{PL'} = \frac{PR}{PQ} = \frac{PR'}{PQ'}.$$

Denote the value of this common ratio by v. Then

$$\frac{KK'}{LL'} = \frac{PK + PK'}{PL + PL'} = v$$

and

$$\frac{QQ'}{RR'} = \frac{PQ + PQ'}{PR + PR'} = \frac{1}{v}.$$

Multiplication yields $KK' \cdot QQ' = LL' \cdot RR'$.

It follows that $(PF + PF')^2 = LL' \cdot RR'$. Since $FF'R'R$ is a parallelogram, $RR' = FF'$. Hence

$$LL' = \frac{(PF + PF')^2}{FF'}$$

has constant length. Since $\angle PFK = \angle PLK = \angle PF'K' = \angle PL'K' = 90°$, $FKLP$ and $F'K'L'P$ are cyclic quadrilaterals. Hence $\angle FLP = \angle FKP = \angle F'K'P = \angle F'L'P$, so that $FF'L'L$ is an isosceles trapezoid. Thus d and d' are situated symmetrically about the ellipse, and since the distance between them is constant, their locations are uniquely determined, independent of the choice of the point P on the ellipse.

From the similar triangles, we have

$$\frac{PF}{PF'} = \frac{PK}{PK'} = \frac{PL}{PL'}.$$

Hence

$$\frac{PF}{PL} = \frac{PF'}{PL'} = \frac{PF + PF'}{LL'}$$

is constant, and the ellipse does satisfies the original definition as well.

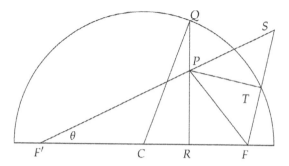

Figure 8. Ellipse based on out new definition.

Let F and F' be fixed points and a a fixed distance. By our new definition, the locus of a point P such that $PF + PF' = 2a$ is an ellipse with foci F and F'. The points on this ellipse can be generated as follows. Take an arbitrary point S satisfying $SF' = 2a$ and let T be the midpoint of SF. Then the point P of intersection of SF' and the line through T perpendicular to SF lies on the ellipse since $PF + PF' = PF' + PS = SF' = 2a$. (See Figure 8.)

Let $FF' = 2\sqrt{a^2 - b^2}$ for a fixed distance $b < a$ and let C be the midpoint of FF'. Let the line through P perpendicular to FF' cut FF' at R, and the circle with center C and radius a at Q. It turns out that we have $QR/PR = a/b$. In other words, the ellipse may be obtained from the circle by contracting the points on the circle toward a diameter by a fixed ratio.

To prove this equality, we establish a coordinate system with C as the origin and the ray CF as the positive x-axis. Then the coordinates of F and F' are $(\pm\sqrt{a^2 - b^2}, 0)$, respectively. Let θ denote $\angle SF'F$. Then the coordinates of S are $(2a\cos\theta - \sqrt{a^2 - b^2}, 2a\sin\theta)$. Hence the coordinates of T are $(a\cos\theta, a\sin\theta)$. The equations of SF' and PT are respectively

$$\frac{y}{x + \sqrt{a^2 - b^2}} = \frac{\sin\theta}{\cos\theta} \quad \text{and} \quad \frac{y - a\sin\theta}{x - a\cos\theta} = \frac{\sqrt{a^2 - b^2} - a\cos\theta}{a\sin\theta}.$$

Solving this system of equation yields the coordinates of P, which are

$$x = \frac{a^2\cos\theta - a\sqrt{a^2 - b^2}}{a - \sqrt{a^2 - b^2}\cos\theta} \quad \text{and} \quad y = \frac{b^2\sin\theta}{a - \sqrt{a^2 - b^2}\cos\theta}.$$

Now

$$QR = \sqrt{QC^2 - RC^2}$$

$$= \sqrt{a^2 - \left(\frac{a^2 \cos\theta - a\sqrt{a^2 - b^2}}{a - \sqrt{a^2 - b^2}\cos\theta} \right)^2}$$

$$= \frac{a}{a - \sqrt{a^2 - b^2}\cos\theta} \left(\left(a - \sqrt{a^2 - b^2}\cos\theta \right)^2 \right.$$
$$\left. - \left(a\cos\theta - \sqrt{a^2 - b^2} \right)^2 \right)$$

$$= \frac{a}{a - \sqrt{a^2 - b^2}\cos\theta} \left(\left(a + \sqrt{a^2 - b^2} \right)(1 - \cos\theta) \right.$$
$$\left. \times \left(a - \sqrt{a^2 - b^2} \right)(1 + \cos\theta) \right)$$

$$= \frac{a}{a - \sqrt{a^2 - b^2}\cos\theta} \left(\left(a^2 - \left(a^2 - b^2 \right) \right)\left(1 - \cos^2\theta \right) \right)$$

$$= \frac{ab\sin\theta}{a - \sqrt{a^2 - b^2}\cos\theta}.$$

Thus we indeed have $QR/PR = a/b$.

Another Definition of the Ellipse

Since an ellipse is a squashed circle, this suggests the following solution to the problem of the cat on the sliding ladder. Let another cat sit on the midpoint C' of a longer ladder $A'B'$ where $A'C' = B'C' = AC = 2BC$. Initially, the height of C above the ground is half that of C'. Let the ladders slide so that they are always parallel to each other. Then $A'C'CA$ is a parallelogram so that CC' is vertical. Let it intersect the floor at a point M. (See Figure 9.) Then triangles MBC and $MB'C'$ are similar, so that $MC' = 2MC$. Since the locus of C' is a quarter of a circle, the locus of C is a quarter of a squashed circle.

To claim that the answer to this problem is a quarter of an ellipse, we must prove the converse of the result at the end of the last section, namely, that a squashed circle is an ellipse.

Let C be the center of a circle with radius a. Let U be a point such that $CU = b$. Let F and F' be points on the diameter of the circle perpendicular to CU, such that $UF = UF' = a$. Let Q be an

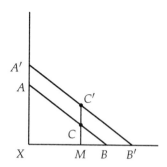

Figure 9. The initial situation, but with two ladders and two cats.

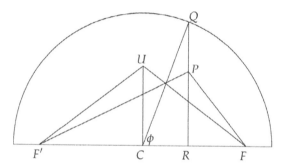

Figure 10. Proving that a squashed circle is an ellipse.

arbitrary point on the circle. Let R be the foot of perpendicular from Q to FF', and let P be the point on QR such that $QR/PR = a/b$. We claim that $PF + PF' = 2a$. (see Figure 10.)

To prove this equality, we establish a coordinate system with C as the origin and the ray CF as the positive x-axis. Then the coordinates of F and F' are $(\pm\sqrt{a^2 - b^2}, 0)$, respectively. Let ϕ denote $\angle QCR$. Then the coordinates of Q are $(a\cos\phi, a\sin\phi)$. Hence the coordinates of P are $(a\cos\phi, b\sin\phi)$. Now

$$(PF + PF')^2$$
$$= \left(\sqrt{(a\cos\phi + \sqrt{a^2 - b^2})^2 + r^2\sin^2\phi} + \sqrt{(a\cos\phi - \sqrt{a^2 - b^2})^2 + r^2\sin^2\phi}\right)^2$$

$$=2a^2\cos^2\phi + 2(a^2 - b^2) + 2b^2\sin^2\phi$$
$$\qquad + 2\sqrt{(a^2\cos^2\phi + a^2 - b^2 + b^2\sin^2\phi)^2 - 4a^2(a^2 - b^2)\cos^2\phi}$$
$$=2a^2\cos^2\phi + 2(a^2 - b^2) + 2b^2\sin^2\phi$$
$$\qquad + 2\sqrt{(a^2\cos^2\phi + a^2 - b^2\cos^2\phi)^2 - 4a^2(a^2\cos^2\phi) + 4a^2(b^2\cos^2\phi)}$$
$$=2a^2\cos^2\phi + 2(a^2 - b^2) + 2b^2\sin^2\phi + 2\sqrt{(-a^2\cos^2\phi + a^2 + b^2\cos^2\phi)^2}$$
$$=2a^2\cos^2\phi + 2(a^2 - b^2) + 2b^2\sin^2\phi + 2a^2\sin^2\phi + 2b^2\cos^2\phi$$
$$=4a^2.$$

It follows that our three definitions of the ellipse are equivalent to one another. A Euclidean establishment of the last equivalence is most desirable.

Bibliography

[1] Martin Gardner. *New Mathematical Diversions*. Washington, DC: Mathematical Association of America, 1995.

[2] V. L. Gutenmacher and N. B. Vasilyev. *Straight Lines and Curves*. Moscow: Mir Press, 1980.

[3] Ed Leonard, Ted Lewis, Andy Liu, and George Tokarsky. "The Parabola: An Euclidean Introduction for Smart Novices." *Mathematics and Informatics Quarterly* 6 (1996), 122–131.

Dances with Tangrams (and without Wolves)

Karl Schaffer

In this article, I will recount experiences with fellow dancers using giant tangrams as props to create dances and explain some of the mathematical problems that have arisen in the choreographing of these dances.

In 1990, Erik Stern and I premiered our first show about dance and mathematics: *Dr. Schaffer and Mr. Stern, Two Guys Dancing About Math.* The show dealt with two characters at odds with each other over whether and how there is math in everything we do, and its dances dealt with rhythm, symmetry, and the play of gravity (á la "a trio for two men and a basketball"). Three years later, in 1993, we were joined by Scott Kim, and the three of us created a new show, *The Secret Life of Squares*, which is more geometric in nature.

In that show we choreographed a dance entitled "The Flying Machine," using oversized tangrams as props. We described the ancient Chinese legends of flying machines, using this probably less ancient Chinese mathematical puzzle. At various points in the dance, we construct seven different squares (Figure 1). Note

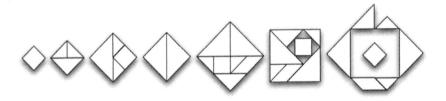

Figure 1. Different squares constructed during "The Flying Machine."

that the parallelogram always appears in only one of its mirror image orientations; this is necessary since the "back side" of each tangram has a handle at its center.

We also make a variety of other forms, some invented for the dance, others well-known. Since the dance is a trio, and there are seven pieces for only six hands, we mounted one of the large tangram triangles on a stand, fixed in the center so that it could rotate freely. The set of tangrams is made of surprisingly sturdy foamcore, and we are still performing with the original set more than ten years later. See [4] for details on construction of giant tangrams.

In 1995 Scott and I, together with Bay Area dancer Barbara Susco, created a third mathematical dance show, *Through the Loop: In Search of the Perfect Square*, in which we explored creations with loops of rope and string. We also did more storytelling with giant tangrams. This is also a trio show. However, the tangram pieces were thick enough that they could be set on the floor without falling over, and we could use one piece to support another, so we did not have to mount a piece on a stand.

Again, we invented a variety of new forms that we used in audience interactions, but not in an actual dance. In this case, we used two different sets of furniture foam tangrams for the audience puzzles, and another much larger set of tangrams made out of "gatorfoam," as a kind of stage set which we periodically move around the stage into new configurations. Gatorfoam is a lightweight, yet sturdy, compressed foam material used for set design—it is like super-compressed foamcore. These pieces have also survived ten years of battering performance.

In 1998 I created a fourth show, *The Day the Numbers Disappeared*, in which we used the seven "Wisdom Plates of Sei Shonagon," a Japanese puzzle also based on the dissection of the square.

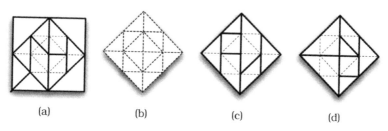

(a)   (b)   (c)   (d)

Figure 2. (a) Butterfly Wing Tables, 1617. (b) Underlying grid? (c) Wisdom Plates of Sei Shonagon, 1742. (d) Tangrams, 1796.

The "seigrams," as we have come to call them, date at least to 1742, fifty years prior to the first known appearance of tangrams in China. But both appear to have the same underlying grid of 16 isosceles right triangles (see Figure 2(b)), which also might be seen to underlie the "Butterfly Wing Tables," a set of tables that could be moved into a variety of configurations depending on the event (see Figure 2(a)). Jerry Slocum points out that the five pieces in the right half of the Butterfly Wing Tables design can be converted into the tangrams with the addition of two line segments dividing two of the pieces [5]. The tables were invented by Ko Shan and published in a book he authored in 1617. An earlier set of rectangular "Banquet Tables" are described in a book published in 1194 by Huang Po-ssu.

The storyline behind *The Day the Numbers Disappeared* is that, one day, all the numbers do just that—"disappear"—and so people cannot even speak about them. We used the seigrams for a four-person dance, entitled "The Torch of Hope," set to a song whose lyrics use nonspecific names for number: a fleet of ships, a billow of clouds, a nest of vipers, and so on. Because we were using four dancers to hold up seven shapes, the only problems with manipulating the shapes were the usual ones in such dances with oversized shapes, such as: Can one person holding two pieces really reach as far as necessary to make a certain figure? What is an elegant way to move from one shape to another? Do pieces need to switch from one hand to another or one dancer to another at any point?

Recently we've toured both of these more recent three- and four-person shows, *Through the Loop* and *The Day the Numbers Disappeared*, to schools and other venues in the San Francisco Bay Area.

But now we have a problem: the four-person show is more difficult to maintain in performance readiness, as dancers are wont
to move off to New York, or drop out to become mortgage bro-
kers, or go back to school. Also, the economics are such that the
four-person show only breaks even or loses a little money, as we
have four people to pay, while the three-person show makes a little
money. We charge the same amount for each show, since otherwise schools would always choose the less expensive show. So I've
been working to reset the four-person concert for three dancers.

But this means redoing all the four-person dances for three people. For example, last year I found a way to redo a PVC-polyhedron
dance so three dancers could make the skeleton of a cube, octahedron, or tetrahedron easily. This year I worked on three-person
sets of seigrams. The goal was to reduce the number of pieces from
seven to six that could still combine to form the same shapes that
we use in the four-person dance. Is it possible to do this and still
make the same (or almost the same) shapes that we made with
seven pieces?

While searching for a solution, I happened on this statement by
Stewart Coffin [1] on dissection puzzles such as tangrams:

> Satisfactory dissection puzzles of this type with fewer
> than seven pieces are not as common, but possible. Con-
> sider the experience of another puzzle acquaintance of
> mine, Bill Trong. Bill made for himself a Tangram set
> from published plans, but he carelessly failed to make
> one cut, so he ended up with two of the pieces joined
> together and thus made a set of six pieces. Surprisingly,
> he found he could construct all 13 of the convex pat-
> terns [that can be made with tangrams] with this set.
> Which two pieces were joined together? Judge for your-
> self if this six-piece version is an improvement over the
> original Tangram.

Finding the 13 convex figures that can be made with a full set
of tangrams is a puzzle in itself. In 1942, Fu Traing Wang and
Chuan-Chih Hsiung proved that no more than these 13 different
convex tangrams can be formed [6]. Their method is first to show
that the 16 congruent right isosceles triangles that compose the
tangrams may themselves be put together to form exactly 20 convex figures (Figure 3). They do this by noting that if the short side
of one of these triangles is rational in length, then the hypotenuse

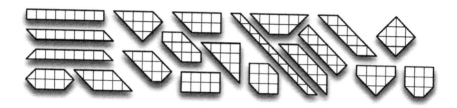

Figure 3. The 20 convex figures constructible with 16 right isosceles triangles.

must be irrational, and then they show that any convex figure construction can only place rational sides together, and similarly for irrational sides. Proving additionally that only 13 of these 20 figures are constructible with the tangrams then involves actually constructing those that are possible and showing that the rest are not possible. This is not too difficult, and it is also an easy puzzle to see that the seigrams will construct 16 of the convex patterns.

All 16 may be constructed if a particular pair of pieces is glued together. If a different pair is glued together, then 13 of the 16 may be constructed. See Figure 4. (I leave the further constructions of the convex figures with either the tangrams or seigrams as puzzles for the reader. The "helping lines" make the task easier.)

It turned out that this latter pairing (Figure 4(b)) allowed nearly all of the shapes from the dance "The Torch of Hope" to be constructed, with minimal changes. In Figures 5 and 6 the "before and after" figures are shown, along with the six modified seigrams used. These six shapes unfortunately no longer tile (that is, cover with no gaps or overlaps) the square, but that is not important for this dance. Also, the two smaller trapezoids are used in mirror image forms (as with our oversized tangrams, the handle on the back makes them actually different in performance). The dance was performed for the first time by only three dancers on February 18, 2005. Several of the figures in Figures 5 and 6 do not use the full set of seigrams, or involve placing the "rational" edges against the "irrational" edges; one of the exaltation of larks figures has the square overlapping another piece.

In honor of Martin Gardner and of the G4G6 theme of the cube, I searched for a way to express the letters M and G using the seigrams, and also for a way to fold them into a cube. Many

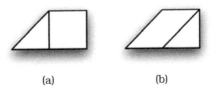

Figure 4. (a) With these two seigrams glued together, 16 convex figures can be constructed. (b) With these two seigrams glued together, 13 convex figures can be constructed.

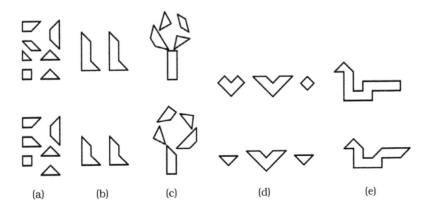

Figure 5. (a) Pieces used. (b) A patter of feet. (c) A bouquet of flowers. (d) A swarm of bees. (e) A nest of vipers.

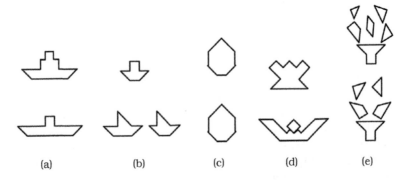

Figure 6. Seven pieces (top) and six pieces (bottom): (a) A fleet of ships, part 1. (b) A fleet of ships, part 2. (c) A drop of rain. (d) An exaltation of larks. (e) A torch of hope.

Figure 7. The seigram alphabet.

Figure 8. The M and G that, when scaled, fold into a cube.

tangram alphabets have been published (e.g., see [2, 3, 5]). The seigrams lend themselves to a nice alphabet (Figure 7), which the reader might like to solve.

If the hypotenuse of the right isosceles triangles making up the seigrams and tangrams has length 1, then the full set of seigrams, like the tangrams, has total area 4. The surface of the unit cube has area 6, however. One way to cover the unit cube with these pieces would be to use two sets, one of area 4 and the other of area 2. I leave it to the reader to figure out how to properly scale either the M or the G in Figure 8 so that the two fold together to tile the unit cube. Each letter may itself be tiled with either the tangrams or the seigrams.

For several final problems, the reader might also like to try the following:

Problem 1. Use these seigrams to make one figure that can be folded to make a cube:

Problem 2. Explain why it is impossible to use these seigrams to make one figure that will fold into a cube:

Problem 3. Finally, make the left half of this wolf with the tangrams and the right half with the seigrams; then switch and make the left half with the seigrams and the right half with the tangrams:

Figure 9 shows the solution to the problem presented in Figure 8, and Figure 10 shows the solution to Problem 3.

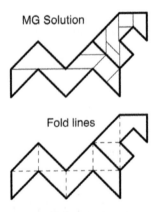

Figure 9. Solution to problem posed in Figure 8.

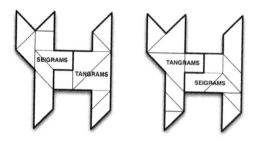

Figure 10. Solutions to Problem 3.

Acknowledgment. The title is based on an idea of Erik Stern for a dance concert: "Dances without Wolves."

Bibliography

[1] Stewart T. Coffin. *The Puzzling World of Polyhedral Dissections*. Oxford, UK: Oxford University Press, 1990. Second edition: *Geometric Puzzle Design*. Wellesley, MA: A K Peters, 2006.

[2] Barbara E. Ford. *The Master Revealed—A Journey with Tangrams*. Vallejo, CA: Tandora's Box Press, 1988.

[3] Ronald C. Read. *Tangrams—330 Puzzles*. New York: Dover Publications, 1965.

[4] Karl Schaffer, Erik Stern, and Scott Kim. *MathDance with Dr. Schaffer and Mr. Stern*. Santa Cruz, CA: Dr. Schaffer and Mr. Stern Dance Ensemble, 2001. (Available at http://www.mathdance.org.)

[5] Jerry Slocum. *The Tangram Book*. New York: Sterling Publishing Co., Inc., 2003.

[6] Fu Traing Wang and Chuan-Chih Hsiung. "A Theorem on the Tangram." *American Mathematical Monthly* 49 (1942), 596–599.

Two Special Polyhedra Among the Regular Toroids

Lajos Szilassi

As is well known, in a regular polyhedron every face has the same number of edges and every vertex is incident at the same number of edges. A polyhedron is called *topologically regular* if further conditions (e.g., on the angles of the faces or edges) are not imposed.

An ordinary polyhedron is called a *toroid* if it is topologically torus-like (i.e., it can be converted to a torus by a continuous deformation), and its faces are simple polygons. A toroid is said to be *regular* if it is topologically regular.

It is easy to see that the regular toroids can be classified into three classes, according to the numbers of edges, vertices, and faces.

There are infinitely many regular toroids in each of these classes, because the number of faces and number of vertices can be arbitrarily large. Hence we study mainly those regular toroids for which the number of faces or vertices is minimal, or those that have other special properties.

Among these polyhedra, we pay special attention to the Császár-polyhedron, which has no diagonal (i.e., every pair of vertices are neighboring), and its dual polyhedron (in a topological sense), the Szilassi-polyhedron, in which every pair of faces are neighboring. The first of these polyhedra was found by Ákos Császár in 1949, and the latter one was found by the author of this paper, in 1977.

The Three Classes of Toroids

Throughout we'll use the fact that the edges of the polyhedra are straight line segments and the faces are planar.

For toroids, Euler's formula $V - E + F = 0$ holds, where V, E, and F are the numbers of vertices, edges, and faces, respectively.

(We could have defined a toroid more generally as an ordinary—but not simple—polyhedron, the surface of which is connected. This generalization will not be needed here.)

Assume that each face of a regular toroid has a edges, and at each vertex exactly b edges meet. Both products $F \times a$ and $V \times b$ are equal to twice the number of edges, since every edge is incident at two faces and two vertices. Hence, from Euler's formula for toroids (above),

$$\frac{2E}{a} + \frac{2E}{b} - E = 0.$$

Since $E > 0$, this leads to the Diophantine equation

$$\frac{1}{a} + \frac{1}{b} - \frac{1}{2} = 0.$$

This equation has only three integer solutions satisfying the conditions $a \geq 3$ and $b \geq 3$. Hence, we can distinguish three classes of regular toroids, according to the numbers of edges incident at each face and each vertex.

As is known, there are only three ways of tiling the plane with regular polygons; namely, with equilateral triangles, squares, and regular hexagons. (Every edge must border exactly two faces; if this condition is omitted the tilings with triangles and squares are not unique.) These three tilings correspond topologically to the three classes

$$
\begin{aligned}
\text{class } \mathbf{T_1}: \quad & a = 3, \quad b = 6, \\
\text{class } \mathbf{T_2}: \quad & a = 4, \quad b = 4, \\
\text{class } \mathbf{T_3}: \quad & a = 6, \quad b = 3.
\end{aligned}
$$

Figure 1. A toroid with 48 triangles in class T_1.

If a *sufficiently large* rectangle is taken from such a tiled plane, and its opposite edges are glued together, we obtain a map drawn on a torus that is topologically regular. (Two opposite edges of the rectangle are first glued together to form a cylindrical tube, and then the remaining two—now circular—edges are glued together, yielding a torus.) If the resulting regular map drawn on the torus has sufficiently many regions, there is no obstacle in principle to its realization with plane surfaces. We may say, therefore, that each of the three classes contains an infinite number of regular toroids. However, it is interesting to determine for each class the smallest number of faces or vertices required to construct a regular toroid in that class, possibly with the restricting condition that the faces or solid angles belong to as few congruence classes as possible.

Class T_1 Toroids

With the use of a sufficiently large number of triangles we can easily construct a toroid belonging to class T_1 (Figure 1). It would be appropriate here to formulate the following problem: what is the minimum number of triangles required to construct a toroid?

An interesting construction in class T_1 requires that the faces of the toroid be colored with seven colors in such a way that all regions of the same color are adjacent to all of the other regions. In this way we can realize with polyhedra Heawood's well-known map drawn on a torus. Any two regions of this map are adjacent. It is even feasible to do this in such a way that regions of the same color are not only adjacent but also congruent (Figure 2).

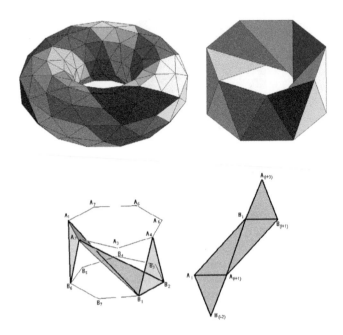

Figure 2. Realization of Heawood's map with a toroid constructed of seven regions, with all of these regions congruent and adjacent to every other region. Every region consists of (for example $4 \times 10 = 40$, or only four) triangles. (See Color Plate V.)

Such a toroid with a minimal number of faces has regions consisting of four triangles that are congruent in pairs. For its construction, let us consider the regular heptagon $A_1, A_2, A_3, \ldots, A_7$.

Let us describe the numerical data for the above construction.

We rotate it by the angle $\frac{5}{2}\frac{2\pi}{7}$ about its center and then shift it in the direction perpendicular to its plane. In this way we obtain the regular heptagon $B_1, B_2, B_3, \ldots, B_7$. For each $i = 1, 2, \ldots, 7$, the figure consisting of the triangles $A_i A_{i+1} B_{i-2}$, $A_i B_i A_{i+1}$, $B_i B_{i+1} A_{i+1}$, and $B_i A_{i+3} B_{i+1}$ is colored one color and is considered as one region (see Figure 2). (If a particular index does not lie between 1 and 7, 7 is either added to it or subtracted from it so as to yield a number between 1 and 7, i.e., indices are taken (modulo 7) + 1.)

If the ith region is rotated by the angle $\frac{2\pi}{7}$ around the axis joining the canters of the two regular heptagons, we obtain the $(i+1)$-

Edges	V_1		V_2		V_3	
$i = 1, 2, \ldots, 7$	d	f	d	f	d	f
$\overline{A_i A_{i+1}}$	6	$64°1'$	6	$51°45'$	6	$43°21'$
$\overline{A_i B_{i-2}}$	6	$150°13'$	8	$152°13'$	10	$153°1'$
$\overline{A_i B_i}$	13.48	$51°12'$	14.48	$65°11'$	13.48	$74°33'$
$\overline{A_{i+1} B_i}$	10.04	$332°13'$	11.35	$325°13'$	10.04	$320°43'$

Table 1. The toroid with seven congruent, pairwise adjacent regions; d = edge length and f = face angles belonging to edges.

th region. These regions are therefore indeed congruent, and together form a toroid. Examining the indices of the edges bordering the regions, one can see that each region is indeed adjacent to all of the others. For example, the neighbor of the ith region along the edge $\overline{A_i B_i}$ is the $(i + 1)$-th region, and its neighbor along the edge $\overline{A_i B_{i-2}}$ is the $(i - 3)$-th region.

For the construction of the polyhedron, we may arbitrarily fix the distance between the planes of the two regular heptagons or—for example—the sides of the isosceles triangle $A_i A_{i+1} B_{i-2}$. From these data, the remaining data may be calculated. Table 1 provides the data for three variants of the toroid, differing in their edge lengths.

This polyhedron is a regular toroid in class T_1. Its faces belong to two congruence classes and its solid angles to one congruence class, i.e., they are congruent. It has $7 \times 4 = 28$ faces.

We can also obtain a toroid of the same kind (i.e., with congruent solid angles and with two types of faces) in class T_2 that has fewer faces and vertices, if we begin the construction with a regular hexagon instead of a heptagon. This toroid has 12 vertices and only 24 faces. A regular toroid with an even smaller number of faces cannot be obtained in this way, because—for instance—for a regular pentagon all of the edges $A_i B_i$, meet in one point, the center of symmetry of the figure. At every vertex of a regular toroid in class T_1 exactly six edges meet, so that such a toroid has at least seven vertices.

The Császár-polyhedron [1, 3, 7] [6, pp. 244–246] is an example of such a toroid with only seven vertices. This polyhedron does indeed belong to class T_1, for any two of its vertices are joined by an edge, and thus six edges meet at each vertex. The number of its vertices is the lowest possible, not only in class T_1. It can readily be seen that a toroid with fewer than seven vertices does not exist.

The toroid denoted by C_0, which is constructed on the basis of data published by Professor Ákos Császár at Budapest University, a member of the Hungarian Academy of Sciences [3], appears fairly crowded.

It has a dihedral angle that is greater than $352°$. We have prepared a computer program to search for a less crowded version. Given the rectangular coordinates of the seven vertices, the program first checks whether the polyhedron defined by the seven points intersects itself. If it does not, the program then calculates the lengths of the edges, the inter-edge angles, and the dihedral angles of the polyhedron. Tables 2 and 3 show the data for five variants of the polyhedron. Variant C_1, can be obtained from C_0 by a slight modification of the coordinates of the vertices, whereas C_1, C_2, C_3, and C_4 are *visually different* from each other.

Let us consider two models of the Császár-polyhedron to be *visually identical* if

- one of the polyhedra can be transformed into the other by gradually changing the coordinates of one of the polyhedra without causing self-intersection in the surface;

- one of the polyhedra can be transformed into the other by reflection;

- one of the polyhedra can be transformed into the other by executing the above operations one after another.

Otherwise, the two Császár-polyhedra are considered to be *visually different*.

Let us try to find a variant among the visually identical polyhedra that is not too crowded and is aesthetically pleasing, and— on the other hand—let us try to find two visually different versions. J. Bokowski and A. Eggert proved in 1986 that the Császár-polyhedron has only four visually different versions [1].

It is to be noted that in topological terms the various versions of the Császár-polyhedron are isomorphic: there is only one way to draw the full graph with seven vertices on the torus (see Figure 3). The vertices of the polyhedra are marked accordingly. In this way, the faces of the polyhedron can be identified with the same triplets of numbers.

(1-2-6)	(1-4-2)	(5-3-2)	(4-1-3)	(2-7-6)	(3-7-2)	(1-7-3)
(6-5-1)	(6-3-5)	(2-4-5)	(3-6-4)	(5-7-1)	(4-7-5)	(6-7-4)

Table 2. Faces of the Császár-polyhedron.

Vertices	C_0			C_1			C_2		
	x	y	z	x	y	z	x	y	z
1	3	-3	0	$4\sqrt{15}$	0	0	12	0	0
2	3	3	1	0	8	4	0	$6\sqrt{2}$	$6\sqrt{2}$
3	1	2	3	-1	2	11	3	-3	$6\sqrt{2}-3$
4	-1	-2	3	1	-2	11	-3	3	$6\sqrt{2}-3$
5	-3	-3	1	0	-8	4	0	$-6\sqrt{2}$	$6\sqrt{2}$
6	-3	3	0	$-4\sqrt{15}$	0	0	-12	0	0
7	0	0	15	0	0	20	0	0	$12\sqrt{2}$

Vertices	C_3			C_4		
	x	y	z	x	y	z
1	12	0	0	12	0	0
2	0	12	$12\sqrt{2}$	0	12	$12\sqrt{2}$
3	-4	-3	$\frac{13\sqrt{2}}{2}$	-3	3	$8\sqrt{2}$
4	4	3	$\frac{13\sqrt{2}}{2}$	3	-3	$8\sqrt{2}$
5	0	-12	$12\sqrt{2}$	0	-12	$12\sqrt{2}$
6	-12	0	0	-12	0	0
7	0	0	$\frac{26\sqrt{2}}{3}$	0	0	$4\sqrt{2}$

Table 3. Coordinates for the four variants of the Császár-polyhedron.

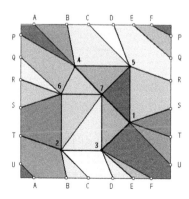

Figure 3. The full graph with seven vertices that can be drawn on the torus.

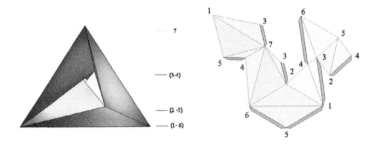

Figure 4. Variant C_1 of the Császár-polyhedron.

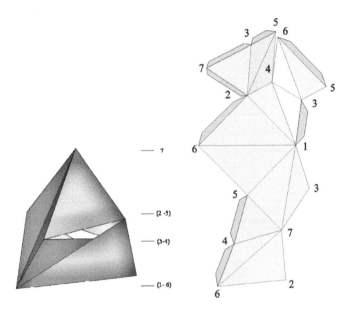

Figure 5. Variant C_2 of the Császár-polyhedron.

It may be observed that in all of the variants, vertices 1 and 6, 2 and 5, and 3 and 4 are reflected images of each other relative to the z-axis of the coordinate system. Accordingly, the pairs of faces in each of the last two columns of Table 2 are congruent. The solid angles corresponding to the foregoing vertex pairs are also congruent. Therefore, in all four variants the faces belong to seven congruence classes, and the solid angles to four congruence

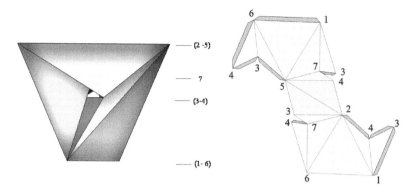

Figure 6. Variant C_3 of the Császár-polyhedron.

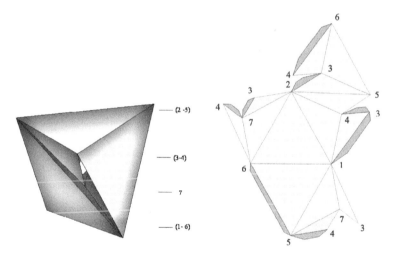

Figure 7. Variant C_4 of the Császár-polyhedron.

classes. For this reason the triangles shown one below another in Table 2 are congruent.

One variant is shown in Figures 4–7 for each of the four versions, which are (in our opinion) not too crowded. The pictures of the four versions are shown together with nets suitable for producing the models. Table 3 includes the coordinates of the vertices, while Table 4 shows the values of the edge lengths and the face angles belonging to the edges.

Edges	C_0		C_1		C_2		C_3		C_4	
	d	f	d	f	d	f	d	f	d	f
$(1-6)$	8.5	153°	31.0	127°	24	90°	24	71°	24	71°
$(2-5)$	8.5	321°	16.0	344°	16.9	270°	24	54°	24	56°
$(3-4)$	4.5	253°	4.5	257°	8.5	114°	10	76°	8.5	286°
$(2-4) = (5-3)$	6.7	78°	12.3	69°	6.9	296°	12.6	204°	16.3	191°
$(2-3) = (5-4)$	3.0	216°	9.3	209°	12.2	35°	17.4	42°	11.0	103°
$(3-7) = (4-7)$	12.2	269°	9.3	279°	12.2	291°	5.9	244°	7.1	22°
$(2-7) = (5-7)$	14.6	18°	17.9	36°	12	61°	12.9	340°	16.5	307°
$(1-5) = (6-2)$	6.1	87°	17.9	90°	16.9	90°	24	53°	24	22°
$(1-2) = (6-5)$	6.1	44°	17.9	67°	16.9	15°	24	51°	24	66°
$(1-4) = (6-3)$	5.1	352°	18.3	343°	16.2	237°	12.6	157°	14.8	39°
$(1-3) = (6-4)$	6.2	58°	19.9	57°	11.0	279°	18.7	339°	19.0	272°
$(1-7) = (6-7)$	15.3	76°	25.3	57°	20.8	24°	17.1	74°	13.3	272°

Table 4. Data for the four variants of the Császár-polyhedron; d = edge length and f = face angles belonging to edges.

Class T_2 Toroids

Class T_2 of regular toroids consists of those torus-like ordinary polyhedra in which four edges meet at each vertex and the faces are quadrilaterals (Figure 8). This type of regular toroid is the easiest to construct.

Let us take a (e.g., regular) p-sided polygon, and rotate it by $(k/q)2\pi$, where q is an integer ≥ 3, and $k = 1, 2, \ldots, q$, about a straight line t that lies in the plane of the polygon but does not intersect it. The resulting toroid, which consists of $p \times q$ trapezoids (or rectangles), is regular, and belongs to class T_2. As an example, for $p = q = 3$ the toroid in Figure 9 is obtained. This is the member of class T_2 with the minimum possible number of faces, since every toroid of type T_2 has at least nine vertices (and nine faces). (Each vertex is incident at four faces, which together have a total of nine vertices. For ordinary polyhedra any two of these nine vertices must be distinct.)

In the case $p = 3$, $q = 4$ (or $p = 4$, $q = 3$), the above procedure yields a regular toroid in T_2 with 12 faces and 12 vertices. However, it is not known whether there exists a regular toroid in T_2 with 10 or 11 faces (vertices), although a graph having 10 or 11 vertices (and regions) and which belongs to the T_2 class can be drawn on the

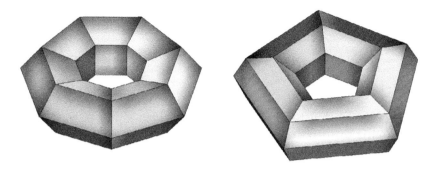

Figure 8. Two toroids with 35 faces in class T_2.

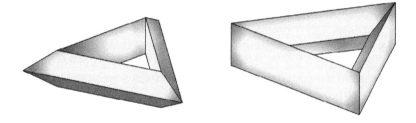

Figure 9. A toroid with minimal faces in class T_2.

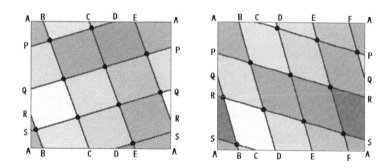

Figure 10. A graph belonging to Class T_2 drawn on a torus. Vertices 10 and 11 have the same number of regions.

torus (see Figure 10). If it does exist, it would have to be obtained
by a different method, since this one requires that F be a product
of two integers each ≥ 3.

Class T_3 Toroids

When constructing toroids belonging to class T_3, care should be
taken to make sure that the six points that define one region (face)
are in the same plane. This requirement can be easily met if the
toroid has a "sufficiently large number" of faces (see Figure 11).

It can be seen that every toroid of Class T_3 has a concave face,
and it is even possible for all of the faces to be concave. An example
of such a toroid, which is shown in Figure 12 (left), consists of 12
L-shaped hexagons. These hexagons are of two types in terms of
congruency.

A somewhat more complicated toroid, shown below, also belongs to Class T_3. It consists of 24 L-shaped hexagons (see Figure
12 (right)). In this case there are four types of hexagons.

We shall prove that there exists in class T_3 a regular toroid with
nine faces such that its faces and solid angles belong to two, respectively three, congruence classes.

Let us project a cube perpendicularly onto a plane π that is perpendicular to one of its internal diagonals that has been selected
in advance. The resulting projection is a regular hexagon. Hence
the projections of any two skew facial diagonals, parallel to π, that

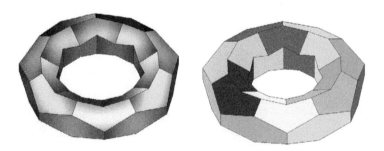

Figure 11. A toroid with 42 faces in class T_3. (The faces of a toroid can
be colored in such a way that identically colored regions composed of six
faces are adjacent and congruent in pairs.)

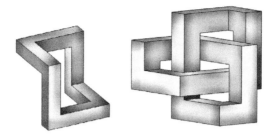

Figure 12. Toroids of Class T_3 consisting of L-shaped hexagons.

are incident at adjacent faces of the cube trisect each other. This means that the line segments joining the corresponding points of trisection of the facial diagonals in question are parallel to the selected internal diagonal.

Utilizing this we can pierce the cube with a triangular prism whose axial edges are parallel to the diagonal of the cube and pass through the points of trisection of the pairs of skew facial diagonals of the cube. The *external* part of the resulting toroid surface consists of the six mutually congruent concave hexagons remaining from the faces of the cube, while its *internal* part is formed by the three mutually congruent convex hexagons that arose during the penetration (see Figure 13).

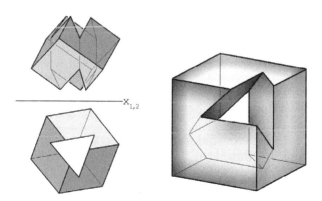

Figure 13. A toroid with nine faces that are of two kinds with respect to their congruency.

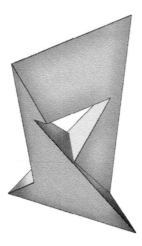

Figure 14. Toroid with seven faces topologically isomorphic with Heawood's seven-color toroidal map.

Seven-Faced Regular Toroid

We have just created a regular toroid consisting of only nine faces. One might wonder: is it possible to create a (regular) toroid from even fewer faces?

As we have seen, the most important property of the Császár-polyhedron is that any two vertices are joined by an edge. There is a very close relationship—a duality—between this polyhedron and the polyhedron with the lowest number of faces in class T_3, the main characteristic of the latter being that any two faces have a common edge (see Figure 14[1]). This relationship is partly topological; however, if we create a new polyhedron by means of a projective transformation from a given polyhedron, e.g., with the use of polarity referring to a sphere, then the new polyhedron will be metric as well [7, 8].

The data for the faces and the net needed to fit them together are given in Figure 15, based on the drawings of Stewart [6, pp. 248–249).

The author of this paper constructed this seven-faced polyhedron in 1977 after producing the dual of the Császár-polyhedron,

---

[1] It was Martin Gardner who first used the term "Szilassi polyhedron" to identify this polyhedron [4].

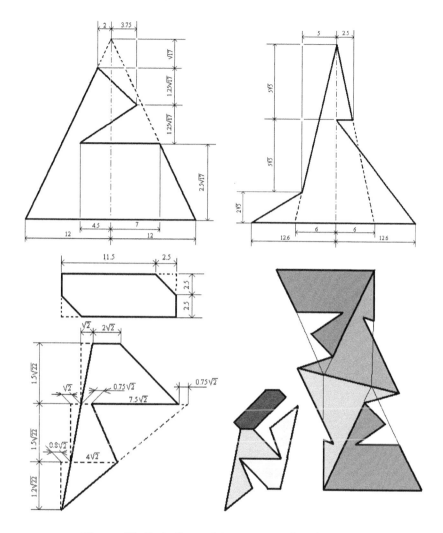

Figure 15. Data for making a seven-faced toroid.

using the spherical polarity of a sphere. The structure obtained in this way, however, consisted of self-intersecting polygons. A computer-assisted analysis was required in order to find the undesirable intersections, as a result of which it was possible to modify the data so as to obtain the above polyhedron (see Figure 15) bordered by simple polygons.

Once one model of the structure is known, it is easy to develop a straightforward method to construct the polyhedron. This will be briefly described below.

- Consider a tetrahedron, which has an axis of symmetry. Assume that this axis is aligned along the z-axis of the coordinate system. The structure will be established in a way that this axial symmetry will be maintained.

- Pierce the tetrahedron with a triangular prism, the edges of which are parallel to the (xy) plane of the coordinate system. Let one of its edges be aligned with two opposite edges of the tetrahedron, and let the other two edges be aligned with the axis of symmetry of two faces. Two quadrangles of the toroid

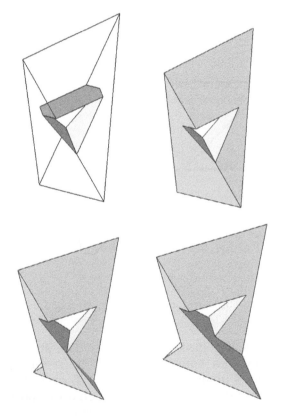

Figure 16. Deduction of the seven-faced toroid.

thus obtained are not yet adjacent to two faces each. For this reason we may complement the structure with a tetrahedron, the face planes of which coincide with the planes of the faces already obtained. In the polyhedron thus produced, any two faces are adjacent. However, two of the faces are not simple hexagons; one of the vertices of each of these hexagons is also on the opposite edge.

- We can eliminate this undesirable coincidence by shifting the edge of the piercing prism (i.e., the edge that intersected two edges of the tetrahedron) closer to the opposite face. Now we obtained the ordinary polyhedron we have been looking for (see Figure 16).

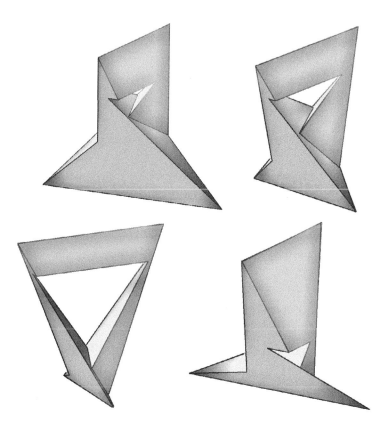

Figure 17. Several (visually identical) variants of the seven-faced toroid.

Starting from the same tetrahedron we can now produce more or less airy (harmonic) variants of the obtained toroid, as shown in Figure 17.

These seven-faced toroids are not regarded as visually different, even though they can be transformed into one another only by reflection in a plane. The author does not know whether this polyhedron has a visually different version (in the manner of the Császár-polyhedron).

Any variant of this polyhedron is also symmetrical about the z-axis of the coordinate system. Thus its faces belong to four congruence classes, and its vertices to seven congruence classes.

In 2005, J. Bokowski proved that the Szilassi-polyhedron—if it has an axis of symmetry—has *only one* visually different version [2, pp. 238–243].

Bibliography

[1] J. Bokowski and A. Eggert. "Toutes les réalisations du tore de Moebius avec sept sommets" (All Realizations of Moebius' Torus with Seven Vertices). *Topologie Struct.* 17 (1991), 59–78.

[2] J. Bokowski. *Computational Oriented Matroids.* Cambridge, UK: Cambridge University Press, 2005.

[3] Á. Császár. "A Polyhedron without Diagonals." *Acta Scientiarum Mathematicarum* 13 (1949–1950), 140–142.

[4] Martin Gardner. "The Minimal Art." *Scientific American* 11 (1978), 22–32.

[5] Martin Gardner. *Fractal Music, Hypercards, and More Mathematical Recreations from Scientific American Magazine.* New York: W. H. Freeman, 1992.

[6] B. M. Stewart. *Adventures Among the Toroids*, revised second edition. Okemos, MI: B. M. Stewart, 1980.

[7] L. Szilassi. "Regular Toroids." *Structural Topology* 13 (1986), 69–80.

[8] L. Szilassi. "Egy poliéder, melynek bármely két lapja szomszédos" (in Hungarian). *A Juhász Gyula Tanárképző Főiskola Tudományos Közleményei* II (1977), 130–139.

Plate I. (See page 34.) *Mysterious Island.*

Plate II. (See page 198.) Coxeter, 1992, with four-triangle model by Odom.

Plate III. (See page 207.) Paper weave.

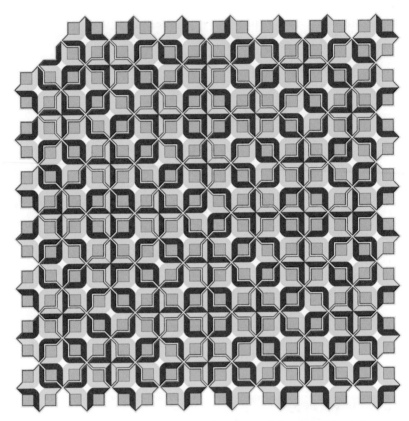

Plate IV. (See page 191.) An aperiodic tiling of the trilobite and cross.

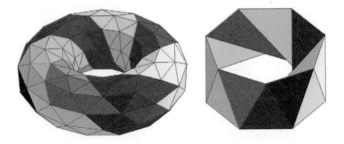

Plate V. (See page 248.) Realization of Heawood's map with a toroid constructed of seven regions, with all of these regions congruent and adjacent to every other region. Every region consists of (for example $4 \times 10 = 40$, or only four) triangles.

Printed and bound by CPI Group (UK) Ltd, Croydon, CR0 4YY

21/10/2024

01777044-0004